David M. Buehlmann
University of Nebraska at Omaha

Dennis Curtin

Applications in Cost Accounting Using Lotus ® 1–2–3®

Cost Accounting

A Managerial Emphasis

EIGHTH EDITION

Charles T. Horngren

Stanford University

George Foster

Stanford University

Srikant M. Datar

Stanford University

Prentice Hall

Englewood Cliffs, New Jersey 07632

NOTE

Lotus 1-2-3 has added a WYSIWYG mode that changes the screen from its original character-based display to a graphic display. Unfortunately, the layout of some items on the screen change when you switch between thses display modes. Since no layout works equally well in both modes, we have chosen to layout the templates on the disk accompanying this text in character-based mode. If some of the text on your screen looks out of alignment, press [/][A][D] (for ***A**dd-In*, ***D**etach*). Highlight the add-in named *WYSIWYG* and press [Enter←]. Press [Q] (for **Q**uit) to return to the worksheet.

A Paramount Communications Company
Englewood Cliffs, New Jersey 07632

Printed in the United States of America

10 9 8 7 6 5 4 3 2 1

ISBN 0-13-184813-5

Prentice-Hall International (UK) Limited, *London*
Prentice-Hall of Australia Pty. Limited, *Sydney*
Prentice-Hall Canada Inc., *Toronto*
Prentice-Hall Hispanoamericana, S.A., *Mexico*
Prentice-Hall of India Private Limited, *New Delhi*
Prentice-Hall of Japan, Inc., *Tokyo*
Simon & Schuster Asia Pte. Ltd., *Singapore*
Editora Prentice-Hall do Brasil, Ltda., *Rio de Janeiro*

CONTENTS

INTRODUCTION

Applications in Cost Accounting with Lotus® 1-2-3® and its accompanying *Cost Accounting Resource Disk* are designed to supplement *Cost Accounting: A Managerial Emphasis* (8th edition) by Charles T. Horngren, George Foster, and Srikant M. Datar. We hope this supplement will enrich your understanding of managerial cost accounting as well as give you an opportunity to learn about the power of the microcomputer as a decision-making tool. To this end, we have provided cases that require the microcomputer, the Lotus® *1-2-3®* spreadsheet program, and the templates on the accompanying *Cost Accounting Resource Disk*. With these tools, you are able to perform a more extensive analysis than would be required by a typical textbook problem. The cases in this supplement require that you interpret and discuss your analysis in a managerial decision-making context. You are asked to define the unique business problem described in the case, to use the templates to analyze data related to alternative courses of action, and finally to make a decision or discuss the implications of the results. We believe these cases represent realistic business situations in a variety of industries.

This supplement does not assume any prior knowledge of microcomputers or 1-2-3. You learn about them in a number of ways as you complete the exercises.

- Exercise 1 takes you step by step through the basics of using a microcomputer and 1-2-3. Exercises 2 and 3 provide additional help if you need it. Each subsequent exercise explains the structure of the template and assumes that you have completed Exercise 1 and are familiar with the program.
- For those of you who want to learn more about creating templates with 1-2-3, optional tutorials are included at the end of each exercise. These tutorials show you, step by step, how to add to, improve on, or make more effective use of the templates. They cover procedures such as entering labels, numbers, and formulas; using data tables; creating, revising, and printing graphs; using windows and fixed titles; using lookup tables; extracting and combining files; and adding items to your own menus. Completing these tutorials will teach you the full power of your program.
- The Appendixes summarize many of the commands discussed throughout this supplement. Refer to them any time you have a question. For example, Appendix A shows you how to make a backup copy of your *Cost Accounting Resource Disk* and introduces you to DOS, your computer's operating system.

Each exercise in this supplement has an accompanying template on the *Cost Accounting Resource Disk*. In general, this supplement provides the following information for each template:

- An overview of the template and its graphs, if any.
- A step-by-step introduction to the first three templates shows how they are organized, how you enter data into them, and how to analyze the results that they calculate for you.
- Several cases that you analyze by entering data into the template.

- Tear-out sheets for you to enter your case solutions for submission to your instructor.
- Optional, step-by-step tutorials that give you experience in template building.

Each exercise assumes that you have read the relevant material in *Cost Accounting: A Managerial Emphasis* (8th edition), solved some of the problems at the end of the chapters, and received appropriate feedback on these problems. If you have, you have learned the basics of a particular managerial accounting technique. You are now ready to explore computerized applications of this technique by solving the cases in this supplement.

Table 1 lists the exercises in this supplement, the chapters in *Cost Accounting: A Managerial Emphasis* (8th edition) to which they are related, and the name of the related template on the accompanying *Cost Accounting Resource Disk*.

Table 1
EXERCISES

Exercise	Text Chapter	Disk Template
1. Getting Acquainted with 1-2-3	None	AUTOLOAN
2. Cost-Volume-Profit Analysis	3	CVP1
3. Multiple Product C-V-P Analysis	3	CVP2
4. Master Budgeting	6	MSTBUDGT
5. Cash Budgeting	6a	CASHFLOW
6. Level 0-3 Variance Analysis	7, 8	VARIANCE
7. Cost Estimation with Regression Analysis	10	COSTBEHV
8. Cost Allocation: Service Dept. Charges	14	SVCDEPT
9. Cost Allocation: Product Overhead Cost	15	PRODCOST
10. Capital Budgeting	20, 21	CAPBUDGT
11. Inventory Management	24	EOQ
12. Managing Division ROI	26	ROI

The AUTO123 File and Menu

The *Cost Accounting Resource Disk* contains a file named AUTO123 that you use to set up templates and retrieve files for exercises. This file should be displayed on the screen automatically when the 1-2-3 program is loaded. It, and all other templates, have user-defined menus that make working with the templates easier. These menus are also automatically displayed in most cases. If your AUTO123 template is not automatically displayed along with its user-defined menu, you can ensure that it does so the next time you load 1-2-3 by changing some default settings as described in Appendix B. If you prefer to load the AUTO123 file and display its menu manually, instructions on doing so are also provided in Appendix B.

EXERCISE 1
GETTING ACQUAINTED WITH 1-2-3

This first exercise introduces you to Lotus 1-2-3 and the templates on the *Cost Accounting Resource Disk* that accompanies this supplement. To complete this exercise, you need the following items:

1. A DOS 2.0 or later version disk (not required if you are working on a hard disk or if your 1-2-3 System Disk is self-booting).
2. A copy of the Lotus 1-2-3 System Disk (not required if you are loading 1-2-3 from a hard disk).
3. A copy of the *Cost Accounting Resource Disk* that accompanies this supplement. This disk contains the templates you use to explore cost accounting. It is ***very*** important that you make a backup copy as described in Appendix A, "Backing Up Your Original Cost Accounting Resource Disk," and use that copy in this and subsequent exercises.
4. If you are not using an Epson (or Epson-compatible) printer, you need to know the setup codes that set your printer to compressed and normal type. These codes are given in your printer's manual or are available from your instructor or lab assistant. Enter them in the spaces provided in Table 2 so that you have them to refer to. You may proceed with this exercise if you do not know your printer codes. You can enter them later in one of the exercises where they are required (Exercises 5, 9, and 10).

Table 2
PRINTER SETUP CODES

Print Mode	Setup String
Compressed type	
Normal type	

UNDERSTANDING SOME KEY TERMS

- When working with spreadsheets, you frequently encounter the terms *worksheet*, *template*, *model*, and *file*. There are subtle differences among these terms.
  - *Worksheet* (or spreadsheet) refers to a spreadsheet's arrangement of blank rows and columns.
  - *Template* refers to labels and formulas entered into cells on the worksheet to perform analysis.
  - *Model* refers to a template that contains data.
  - *File* refers to a model or template stored on a disk or in the computer's memory.

  For simplicity's sake, we use the term *worksheet* for the blank arrangement of rows and columns, and *template* for all forms of data entered on the worksheet.
- You are frequently asked to press the **Enter** (or **Return**) key to enter data and make selections from menus. On the IBM PC, it is the key labeled ↵.

Step 1 Get Ready. Before proceeding:

- On a floppy disk system, insert the DOS disk into drive A. (Insert the 1-2-3 System Disk if it is self-booting.)
- On a hard disk system, open the door to drive A, or eject the disk in that drive.

Note: On a computer connected to a network, your loading procedure may be different.

Step 2 Load the Operating System. To load the operating system, turn on the computer. The on/off switch is usually located on the front of the computer or on the side toward the rear. When you turn the computer on, it runs its internal diagnostic program. In a few moments, the computer may beep, and then drive A spins and its light comes on while the operating system is loaded. If there is no disk in drive A, the computer looks to drive C if the system contains a hard disk drive. If nothing appears on your screen, your display screen may not be on. On some systems, the display screen has a separate on/off switch.

If your computer does not have a clock that is set automatically, in a moment the prompt reads *Enter new date:*. (If you make a typo when entering any commands, press **Backspace** to delete the incorrect characters, and then type them in correctly before pressing **Enter**.)

1. Enter the date as mm-dd-yy (for example, type **01-30-92** for January 30, 1992), and then press **Enter**. The prompt then reads *Enter new time:*.
2. Enter the time as hour:minute (for example, type **1:30**), and then press **Enter**.

Result. The command prompt (usually *A>* on a floppy disk system or *C:\>* on a hard disk system) appears on the screen.

Step 3 Load 1-2-3. If 1-2-3 is not loaded automatically, follow these steps:

- If you are working on a floppy disk system
  - Insert the Lotus disk labeled *System Disk* into drive A.
  - If the command prompt is not *A>*, type **A:** and then press **Enter** to change the default drive to drive A.
  - Insert the *Cost Accounting Resource Disk* into drive B.
- If you are working on a hard disk system
  - Insert the Cost Accounting Resource Disk into drive A.
  - Type **C:** and then press **Enter** to change the default drive to the drive containing the 1-2-3 program files.
  - Use the CD command to change to the directory the 1-2-3 program files are stored in — usually named 123 or LOTUS. For example, to change to the directory named 123, type **CD\123** and then press **Enter**.

To load 1-2-3

Type **123** and then press **Enter**

Result. The drive spins as the program is loaded. In a few moments, the copyright screen appears followed, a few moments later, by the worksheet. When you load 1-2-3, the file AUTO123 (Figure 1) should automatically appear on the screen. If this file or its menu is not displayed, refer to Appendix B, "Using the AUTO123 Template Menu," for instructions on how to retrieve it and display its menu manually or on how to change your program's settings so that these things are done automatically.

Step 4 Use the AUTO123 File. You use AUTO123 to enter your name so that it appears on your printouts, to set up your printer, to retrieve templates, and to quit the program when you have finished. You will explore most of the menu choices in this exercise, and all of them are summarized for easy reference in Appendix B, "Using the AUTO123 Template Menus."

FIGURE 1 The AUTO123 File. When you load 1-2-3 with the *Cost Accounting Resource Disk* in one of the drives, the AUTO123 file is automatically loaded, and its menu is displayed. You can also retrieve AUTO123 just like any other file and display the menu by pressing **Alt-M**.

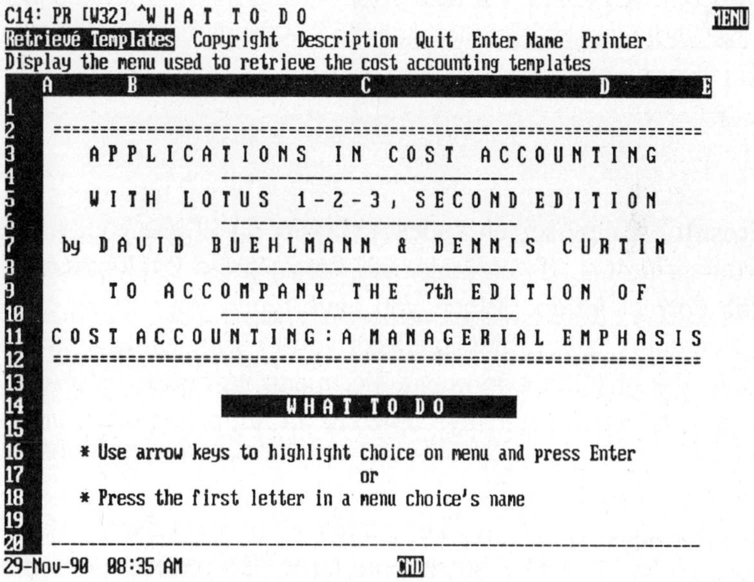

A list of choices called a menu is displayed at the top of the screen. The first choice on the menu, Retrieve Templates, is currently highlighted. This bright highlight is called the menu pointer.

A brief description of the highlighted choice appears on the line below the menu. Press → to move the menu pointer to the next choice. The description on the line below changes. Press **End** to highlight the last choice on the menu, and then press →. The menu pointer wraps around from the last choice to the first choice. You can make selections from the menu by highlighting a choice and then pressing **Enter**. You can also press the first character in the choice's name — the approach used in the instructions throughout this supplement. Let's use the menu to display information, and then enter your name and your printer's setup codes.

To use AUTO123's template menu

Press	**C** (for *Copyright*) to display the copyright page
Press	**D** (for *Description*) to display a description of the disk
Press	**E** (for *Enter Name*) to enter your name

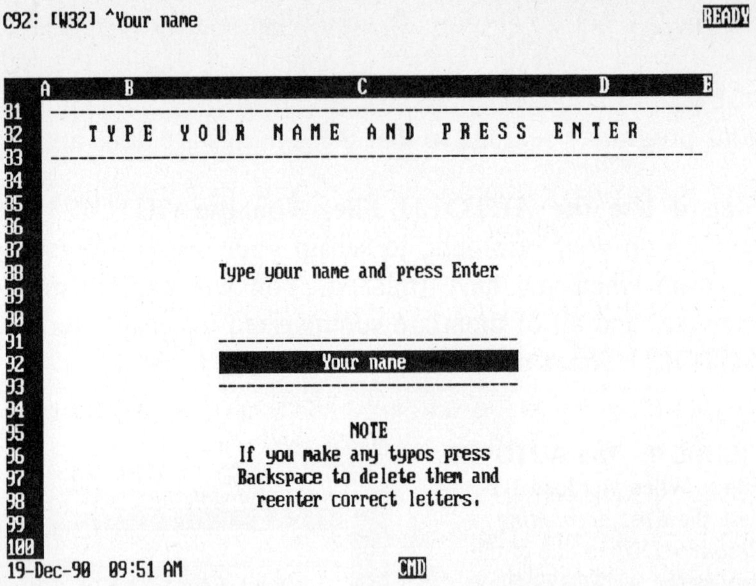

FIGURE 2 The Enter Name Screen. When you press **E** (for *Enter name*), a screen is displayed where you enter your name. Your name is then automatically carried to all the template files that you work with and is printed on your printouts.

Result. A new screen appears (Figure 2). Type your name as you want it to appear on your printouts. If you make any typos, press **Backspace** to delete them, and then reenter the correct letters. When you have typed your name correctly, press **Enter**. The disk drive spins, and a message asks you to wait while a file containing your name is saved onto the disk. In a moment, the menu reappears. You do not have to enter your name again. It is automatically copied to all templates when you retrieve them in the following exercises, and it appears on all printouts of those templates.

Step 5 Printer Setup. The templates in Exercises 5 (CASHFLOW), 9 (PRODCOST), and 10 (CAPBUDGT) are quite large. To reduce the number of pages required to make printouts of these templates, they can be printed in compressed type. To do so, you enter setup codes that set your printer to compressed type and then return it to normal type when printing is completed. Your printer setup codes should be listed in Table 2.

- If you use an Epson dot-matrix printer, the codes have already been entered for you, so you needn't complete this step. These Epson codes (\015 for compressed type and \018 for normal type) are the same for some other printers if they are compatible with Epsons. If you find that your printer uses these same codes, you do not have to enter them. Proceed to Step 6.
- If you don't know what your printer setup codes are, proceed to Step 6. You can enter the codes later by retrieving AUTO123 again, or you can enter them directly into one of the templates where they are required when you get to Exercises 5, 9, or 10.

To set up your printer

Press	P (for *Printer*), and a prompt at the top of the screen reads *Enter setup code for compressed type:*
Type	the setup code for your printer's compressed type, and then press **Enter**. The prompt reads *Enter setup code for normal type:*

Type the setup code for your printer's normal type, and then press **Enter**

Result. The disk drive spins, and a message asks you to wait while files containing the codes are saved onto the disk. In a moment, the AUTO123 template menu reappears.

Step 6 Retrieve the AUTOLOAN Template. The *Cost Accounting Resource Disk* contains a number of templates that you use to explore cost accounting. To use these templates, you must first retrieve them from the disk into the computer's memory. You do this with AUTO123's Retrieve Templates menu.

To retrieve the AUTOLOAN template

Press **R** (for *Retrieve Templates*) to display a list of templates
Press **1** (for *1-AUTOLOAN*)

Result. The drive spins as the file containing the AUTOLOAN template is retrieved. A title page appears along with a message to wait while the program copies the name you entered in Step 5. In a moment, the drive stops, and the message *Press Enter to continue* is highlighted. Press **Enter** to display the AUTOLOAN template.

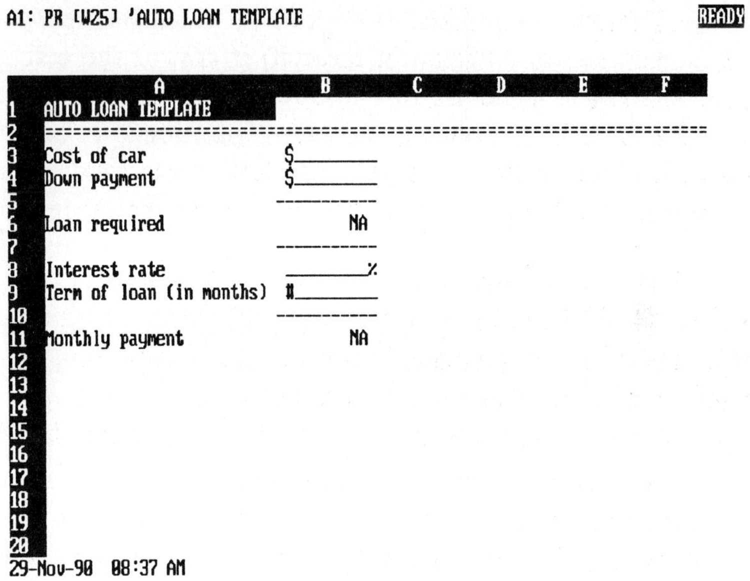

FIGURE 3 The AUTOLOAN Template. When you first retrieve the AUTOLOAN template, it contains blank data entry spaces, but all the labels and formulas have been entered.

Result. The template appears on the screen (Figure 3). Some cells contain data entry spaces, $_____, #______, or ______%. These indicate where you should enter data into the template. On many displays, you also see some cells in a different intensity or color. These cells are unprotected so that you can enter data into them. (On all displays, a *U* is displayed on the first line of the control panel when the cell pointer is in an unprotected cell.) If you try to enter data into other cells, the computer beeps when you press **Enter**, and you have to press **Esc** to continue.

Step 7 Explore Your Program's Screen Display. 1-2-3's screen display has three areas: the control panel, the working area, and the status line (Figure 4).

FIGURE 4 The 1-2-3 Command Menu. When you press the slash (/) key, the command menu is displayed.

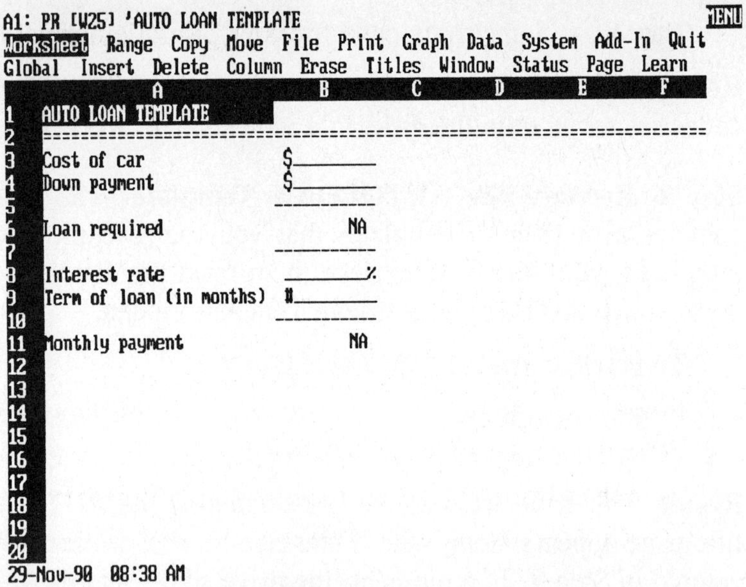

- The control panel is located at the top of the screen, and it has three lines:
  - The first line indicates the cell in which the cell pointer is positioned and the contents and format of the cell, if any. If the cell pointer is in cell A1 (press **Home** if it isn't), this line reads *A1: PR [W25] 'AUTO LOAN TEMPLATE*, indicating that the cell pointer is positioned in column A on row 1, the cell is protected (PR), it is 25 characters wide, and the cell contains the indicated label. At the far right end of the line is the mode indicator, which tells you the status of the program. At the moment, it should read *READY*. Press the arrow keys to move the bright highlight to one of the cells with a data entry space. You will see the letter U on the first line of the control panel. This means the cell is unprotected so that you can enter data into it.
  - The second line displays data as you type it in until you press **Enter** to enter it into the cell or until you press **Edit (F2** on the IBM PC) to edit it. A menu is also displayed on this line when you press the slash key (/) to begin a command.
  - The third line displays a brief description of the highlighted menu selection when the menu is displayed. If choosing a command displays a submenu, the submenu's commands are listed instead of a description.

To display 1-2-3's command menu, press slash (/). When you do so, the mode indicator at the right end of the status line reads MENU to indicate that you are in the program's command menus. Menu choices appear on the second line. Press → to move the menu pointer along the list of choices. As you do so, the line below displays detailed descriptions about each menu item or a list of commands on a submenu. If you press **Enter**, the highlighted choice is executed. To see this, highlight Worksheet and then press **Enter** to display a submenu. You'll learn more about using menus as you proceed through this book. For now, press **Esc** twice to remove the menus from the screen so that the mode indicator reads *READY*.

- The working area is where you enter data and create models. This area contains the following elements:
  - The working area's border is labeled with column letters across the top of the screen and row numbers down the left side.
  - The worksheet area within the border is divided into *cells*, each of which falls at the intersection of a column and a row. The letters and numbers on the border identify each cell. Cell A2 falls at the intersection of column A and row 2, cell B4 falls at the intersection of column B and row 4, and so on.
  - The bright highlight in the worksheet area is called the cell pointer. You move this cell pointer with the arrow keys to point to cells in which you want to enter or edit data.
- The status line displays information as you execute commands. For example, indicators warn you when **Num-Lock**, **Scroll-Lock**, or **Caps-Lock** are engaged. To see these indicators, press each of these keys once to turn their indicator on and then again to turn it off. Before continuing, be sure all these indicators are off. If any are on, press the appropriate key to turn them off.

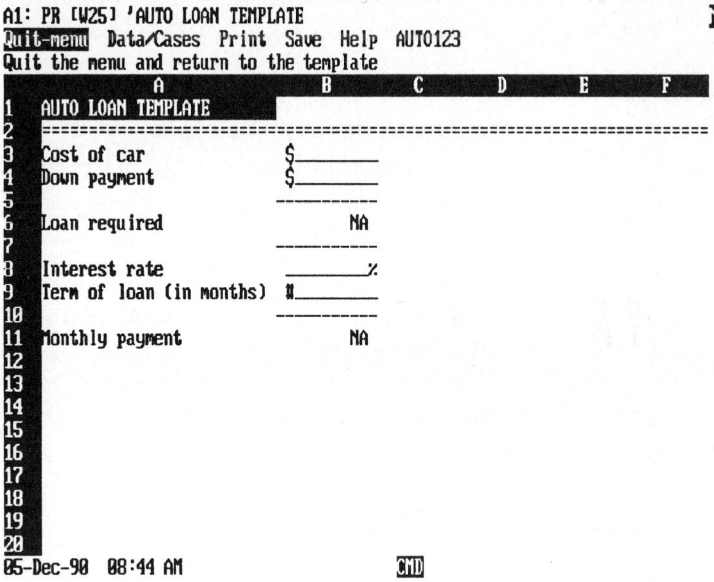

FIGURE 5 The AUTOLOAN Template's Menu. When you press **Alt-M** with this (or any other template) on the screen, the user-defined menu is displayed. This menu lists commands you can execute.

Step 8 Enter Original Numbers. Now let's enter some numbers to use as beginning variables. Begin by assuming you want to purchase a car that costs $10,000 and you have $2,000 as a down payment. The bank has told you they will give you a loan for the $8,000 balance, payable in 36 months at 13%. These original numbers can be entered automatically using the template menu. (In the instructions below, you are asked to press **Alt-M** to display the template's menu. This is shorthand for "hold down **Alt** while you press the letter **M**.")

To enter the original numbers

Press	**Alt-M** to display the template's main menu (Figure 5)
Press	**D** (for *Data/Cases*) to display the Data/Cases menu

Press O (for *Original*) to enter the original numbers

FIGURE 6 AUTOLOAN with Data Entered. When you make a choice from the Data/Cases menu, numbers are automatically entered into your template.

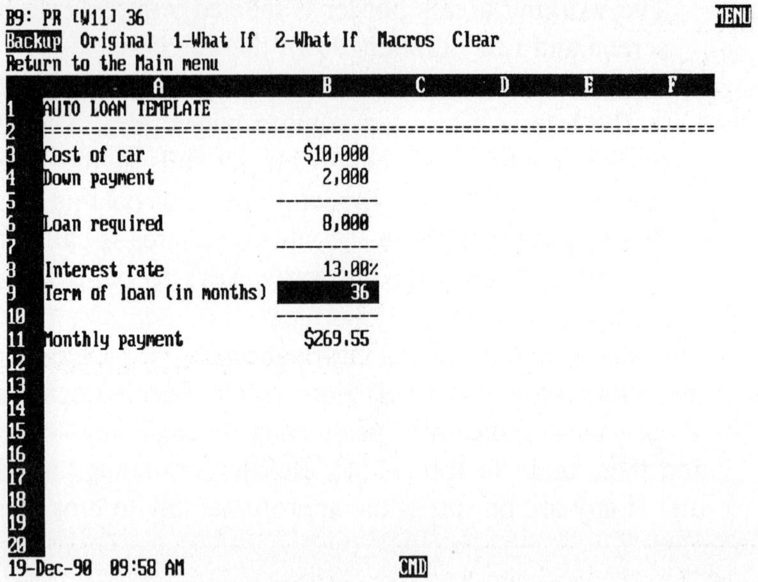

Result. Numbers are automatically entered into the template (Figure 6). The template menu remains on the screen in anticipation of your making additional choices. The original numbers result in a monthly payment of $269.55 (cell B11).

Step 9 Understanding the Menu. Now that you have used the template menu, let's take a moment to see how it works. All the templates on this disk have these menus. Unlike the program's command menus, template menus were written by the authors of this supplement to make the templates easier to use.

To display a macro

Press M (for *Macros*) from the Data/Cases menu

Result. A macro appears on the screen (Figure 7). It is the one you just used to enter original numbers. Macro is a computer term that refers to recording or storing keystrokes so that you can play them back automatically. This macro looks complicated, but you'll be surprised at how simple it is. The first line appears as one of the choices on the menu when you press **Alt-M**. The second line is the description that appears when you highlight the menu choice. The next four lines are simply keys and data that you could have typed from the keyboard. Let's look at the third line (cell H27) to see how it works.

You could enter the same data by pressing the keys listed in the macro. The next three lines, like the third, tell the program to go to other cells and enter data. The final line is a command that displays the menu again when the macro is finished entering data. After studying the macro, press **Enter** to continue.

FIGURE 7 The Macro. The macro for the AUTOLOAN template displays the user-defined menu and contains the commands that are executed when you make a choice from the menu.

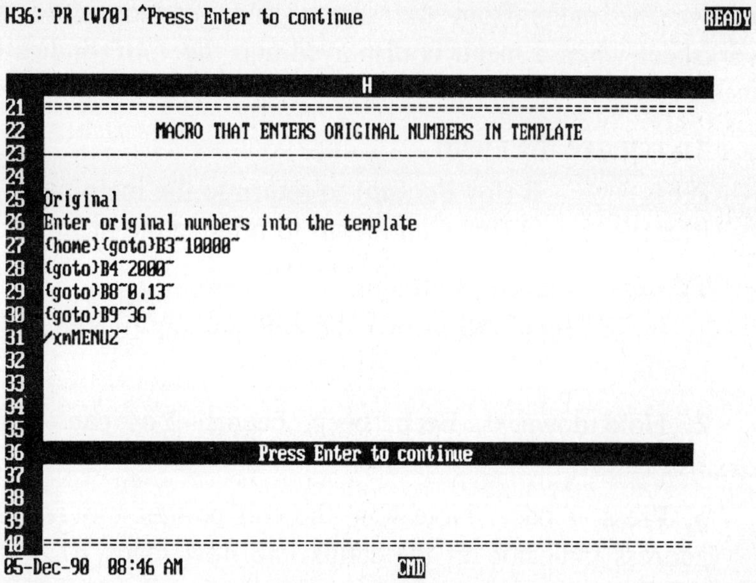

- *{home}* tells the program to press **Home** to move the cell pointer to cell A1.
- *{goto}* tells the program to press **Goto** (**F5** on the IBM PC). If you press this key, a prompt asks you to enter a cell address to go to.
- *B3* is the answer to the prompt.
- ~ (called a tilde) is the symbol that tells the program to press **Enter.**
- *10000* is the amount typed into cell B3.
- ~ presses **Enter** to complete the entry in cell B3.

Step 10 Explore a What-If Using the Menu. Let's explore a what-if to see how the monthly payments change when you increase the period of the loan from 36 to 48 months.

To change the term of the loan

Press 1 (for *1-What If*) to change the term of the loan

Result. The new term of the loan is automatically inserted, and the result is displayed. The monthly payment falls to $214.62.

Step 11 Explore Another What-If. What if you also increase the down payment from $2,000 to $2,500?

To change the down payment

Press 2 (for *2-What If*) to increase the down payment

Result. The new down payment is automatically inserted, and the results are displayed. The loan required decreases from $8,000 to $7,500, and the monthly payment decreases still further to $201.21.

Step 12 Explore the Worksheet. Now that you are familiar with the template menu, practice moving the cell pointer around the worksheet. To do this, you have to first remove the menu from the screen. You cannot move the cell pointer around the worksheet when a menu is displayed and the *CMD* indicator is displayed on the status line.

To remove the menu

Press B (for *Backup*) to return to the main menu
Press Q (for *Quit-menu*) to remove the menu and the *CMD* indicator

To move the cell pointer

1. Press **Home**. Note that the cell pointer immediately returns to cell A1, called home.

2. Hold down ←. Beep, beep, beep!!! You can't move the cell pointer off the worksheet.

3. Press → once. Note how the cell pointer moves one cell to the right. The cell address indicator on the status line now reads *B1:*, the coordinates of the cell in which the cell pointer is currently positioned.

4. Press ↓ three times. Note that the cell pointer moves down one cell at a time. The cell address indicator has now changed to *B4*.

5. Hold down →. Note how it automatically repeats. After the cell pointer reaches the right edge of the screen, the screen begins to scroll. (You may see more macros, but now you know what they are.) The column letters in the border area begin to change when you scroll past the right edge of the screen.

6. Hold down ↓. See what it does?

7. Lost? Just press **Home**. It always returns you to the top left corner of the worksheet.

8. Press **PgDn**. It moves you a screen at a time. You can page quickly through a template this way. Press **PgUp**. It does the same thing but in the opposite direction.

9. Press **Tab**. You move one full screen to the right. Now hold down **Shift** and then press **Tab**. You move left one full screen.

10. Press **Goto** (**F5** on the IBM PC), and a prompt on the second line of the control panel asks you to enter the cell address to go to. Type `M30` and then press **Enter**. The cell pointer immediately jumps to cell M30.

11. Press **Goto** (**F5** on the IBM PC) again. Type `HELP` and then press **Enter**. The cell pointer jumps to the help screen. It does this because we have assigned a name to this screen. You can use the **Goto** key to go to names as well as cell addresses. Press **Goto** again, type `PART 1` and then press **Enter**. Although this template has only one part, the cell pointer immediately jumps to cell A1 because that cell has been named Part 1. All the parts of the templates on the *Cost Accounting Resource Disk* are named so that you can use this command to get to them.

12. Before proceeding, you might want to experiment with some of the other cell pointer movement keys summarized in Appendix D, "Getting Around a Template".

13. When you have finished experimenting, press **Home** so that the cell pointer returns to cell A1.

Step 13 Explore a What-If Manually. The cell pointer should now be in cell A1. Let's explore what-ifs by entering numbers directly into cells of the template. For example, what if you lower the amount paid for the car from $10,000 to $9,000?

- In the instructions throughout this supplement, the heading "GoTo" tells you to move the cell pointer to the indicated cell. Use the arrow keys to move the cell pointer around the screen. With the cell pointer in the indicated cell, type the data shown in the instructions, and then press **Enter** to complete the entry. If the cell already contains data but isn't protected, the new number replaces the old when you press **Enter**.
- If you make any mistakes while entering data, see Appendix E, "Entering Data into a Template."

To change the cost of the car

GoTo B3, type **`9000`** and then press **Enter**

Result. The loan required falls from $7,500 to $6,500, and the monthly payment drops to $174.38.

Step 14 A Final What-If. What if you can lower the interest rate from 13% to 12%?

To change the interest rate

GoTo B8, type **`.12`** and then press **Enter**

Result. The monthly payment drops to $171.17. If you want to keep trying what-ifs, please do so.

Step 15 Explore the Template's Formulas. Now that you have some experience with the template, let's take a moment to see how it works. Move the cell pointer to cell B6, and the formula in the cell is displayed on the first line of the control panel. It reads as follows:

B6: (,0) PR [W11] @IF(B3>0,B3-B4,@NA)

- *B6* indicates the cell pointer's position.
- *(,0)* indicates the cell's format — commas between thousands and zero decimal places.
- *PR* indicates the cell has been protected, so you cannot enter data into it until you turn protection off.
- *[W11]* indicates that the column's width has been changed to 11 characters wide.
- *@IF(B3>0,B3-B4,@NA)* is the formula. The essential component of the formula is *B3-B4,* which subtracts the down payment entered in cell B4 from the cost of the car entered in cell B3. The formula has been nested in an IF statement, so the symbol *NA* is displayed in the cell if the cell where you enter the cost of the car is empty. The formula reads as follows in English: "If the value in cell B3 is greater

than 0, subtract the number in cell B4 from the number in cell B3: otherwise, display *NA* (for Not Available) in the cell."

Move the cell pointer to B11 to display the formula in that cell. The status line displays:

B11: (C2) [PR] [W11] @IF(B8>0#AND#B9>0,@PMT(B6,B8/12,B9),@NA)

- *B11* indicates the cell pointer's position.
- *(C2)* indicates the cells format – currency (dollar signs) and two decimal places.
- *PR* indicates the cell has been protected, so you cannot enter data into it until you turn protection off.
- *[W11]* indicates that the column's width has been changed to 11 characters wide.
- *@PMT(B6,B8/12,B9)* is the function that calculates a monthly payment given three variables. The three variables are the amount of the loan (in cell B6), the annual interest rate (in cell B8) divided by 12 (to calculate the monthly interest rate), and the term of the loan (in cell B9). Instead of entering these variables directly into the function, this function refers to other cells on the template where these variables are entered. This way you can change the variables easily to explore what-ifs. Any numbers you enter into cells to which the function refers are calculated just as if you had entered them into the function directly. This function is nested in an IF statement, so it calculates only when there are values in both cells B8 and B9. If either of these values is zero, the IF statement displays an *NA* in the cell.

Step 16 Save Your Work. When you began this chapter you retrieved the AUTOLOAN template from the disk. When you now save the template, you overwrite the template's previous file on the disk. The next time you retrieve the template it will contain the same numbers it contained when you saved it. In later exercises you may want to save your data in this way if you want to quit before you have completed an analysis. If you do so, you can retrieve it later with the saved data. Before you save a template, be sure the *Cost Accounting Resource Disk* is still in the drive from which you originally retrieved the template.

To save the template

Press	**Alt-M** to display the template menu
Press	**S** (for *Save*)

Result. A message asks you to wait while the program takes a few seconds to save the template. When the drive stops and the highlight jumps to *Press Enter to continue*, a copy of the template with your data has been saved onto the disk. Press **Enter** to return to the template and its menu.

Step 17 Print the Template. Begin by turning on the printer and checking that it has paper in it.

To print the template

Press	**P** (for *Print*)

Result. The printer prints out the template. If you entered your name in the AUTO123 file, it is printed at the top of the page. If the computer's clock is set correctly, the correct date is also printed at the top of the page.

Step 18 Clear Numbers from the Template. The template can be returned to its original form at any time using the Clear choice on the template menu.

To clear numbers from the template

Press	**D** (for *Data/Cases*)
Press	**C** (for *Clear*)

Result. A warning appears telling you that you will lose any work if you haven't saved it. Press **Enter** to continue. (You have two choices: cancel the command or continue.) The numbers are cleared, and blank data entry spaces are automatically inserted in their place.

Step 19 Return to AUTO123. When you have finished with a template, you return to AUTO123 either to retrieve another template or to quit the program if you have finished for the day. When using this command, be sure to save your work first if you want to keep it.

To return to AUTO123

Press	**B** (for *Backup*) to return to the main menu
Press	**A** (for *AUTO123*)

Result. A warning appears telling you that you will lose any work you haven't saved. Here, press **Enter** to continue although you could also cancel the command and return to where you were before you selected the AUTO123 command.

Step 20 Continue or Quit. You can continue by retrieving a new template or entering your own. When you are through for the day, you exit the program and return to the operating system.

To quit the program

Press	**Q** (for *Quit*)

Result. The operating system appears on the screen.

- For more on retrieving a template, see Appendix C, "Retrieving a Template."
- For more on quitting, see Appendix K, "Quitting and Clearing the Screen."

TUTORIAL 1 IMPROVING THE AUTOLOAN TEMPLATE

You can expand on, or improve, any template as you gain experience with it, and the AUTOLOAN template is no exception. One improvement would be to have the template calculate the monthly payments for a range of loans instead of having to enter each data set individually. This would enable you to see at a glance the size of the loan you could afford. Let's see how you would do this. This tutorial assumes that the template AUTOLOAN is displayed on the screen. If it isn't, refer to Exercise 1 for step-by-step instructions or to Appendix C, "Retrieving a Template." When you have finished this tutorial, your template will look like Figure 8.

FIGURE 8 The AUTOLOAN Template with a Range of Payments. After completing this tutorial, your template should look like this figure.

C3: READY

	A	B	C	D	E	F
1	AUTO LOAN TEMPLATE			Loan	Payment	
2	========================	==========	========	========	========	========
3	Cost of car	$10,000		$1,000	$33.69	
4	Down payment	2,000		$2,000	$67.39	
5		----------		$3,000	$101.08	
6	Loan required	8,000		$4,000	$134.78	
7		----------		$5,000	$168.47	
8	Interest rate	13.00%		$6,000	$202.16	
9	Term of loan (in months)	36		$7,000	$235.86	
10		----------		$8,000	$269.55	
11	Monthly payment	$269.55		$9,000	$303.25	
12				$10,000	$336.94	
13				$11,000	$370.63	
14				$12,000	$404.33	
15				$13,000	$438.02	
16				$14,000	$471.72	
17				$15,000	$505.41	
18				$16,000	$539.10	
19				$17,000	$572.80	
20				$18,000	$606.49	

05-Dec-90 08:47 AM

Step 1 Enter Original Numbers. Before entering new data, reenter the original numbers in the template so that your results match ours.

To reenter original numbers

Press **Alt-M** to display the template menu
Press **D** (for *Data/Cases*)
Press **O** (for *Original*)
Press **B** (for *Backup*)
Press **Q** (for *Quit-menu*)

Step 2 Turn Off Protection. Remove cell protection so that you can enter data into any cells on the worksheet.

To turn off protection

Press **/WGPD** (for *Worksheet, Global, Protection, Disable*)

Step 3 Enter Labels. Here you enter labels that identify the two columns into which you will enter data. When entering the labels, you'll press the arrow key to simultaneously enter the label and move the cell pointer to the next cell. After entering the labels, you'll align them in their cells.

To enter labels

Goto	D1, type **`Loan`** and then press → to move the cell pointer to cell E1
Type	**`Payment`** and then press **Enter**

To align the labels

Press	**/RLR** (for *Range*, *Label-Prefix*, *Right*), and the prompt asks you to enter the range of the labels to be aligned
Type	**`D1.E1`** and then press **Enter**

Step 4 Enter Numbers That Represent Possible Loan Amounts. Here you enter one number. Then you enter and copy a formula to automatically generate the rest of the numbers you need. When you enter the formula, you *point* to the cell to which it refers instead of typing its address.

To enter a number and the formula

Goto	D3, type **`1000`** and then press ↓ to move the cell pointer to cell D4
Type	**`1000+`**
Press	↑ to point to cell D3, and then press **Enter**

Result. The cell pointer remains in cell D4, and the formula calculates and displays *2000*. The formula itself, *1000+D3*, is displayed on the first line of the control panel when the cell pointer is in cell D4.

To copy the formula

Press	**/C** (for *Copy*), and the prompt asks you to enter the range to copy FROM
Press	**Enter**, and the prompt asks you to enter the range to copy TO
Type	**`D5.D20`** and then press **Enter**

Result. The range of cells in column D fills with numbers from 1,000 to 18,000. If your results are different, repeat the procedure.

Step 5 Enter a Function. Now, enter a function that calculates the monthly payment for each of the possible loan payments. Functions are small programs built into 1-2-3 to perform complicated mathematical operations.

To enter a function to calculate monthly payments

Goto	E3, type **`@PMT(D3,B$8/12,B$9)`** and then press **Enter**. (The cell pointer remains in cell E3.)

To copy the function

Press	**/C** (for *Copy*), and the prompt asks you to enter the range to copy FROM

Press	**Enter**, and the prompt asks you to enter the range to copy TO
Type	`E4.E20` and then press **Enter**

Result. The range of cells in column E fills with numbers from 34 to 606 (Figure 8). If your results are different, repeat the procedure, being especially careful when entering the payment function and the dollar signs. These dollar signs are entered so that the function works when copied to other rows. They fix the reference to the row numbers so that when you copy the function it always refers to the rows on which interest (row 8) and period (row 9) are entered in column B. A cell reference in a formula that continues to refer to the same cell when the formula is copied is called an absolute reference.

Step 6 Format the Numbers to Display Them as Currency. Here you first format all the numbers with zero decimal places and then change the format of the payments to two decimal places.

To format the loan

Press	**/RFC** (for *Range*, *Format*, *Currency*), and the prompt asks for the number of decimal places
Type	`0` and then press **Enter**. The prompt asks for the range to format
Type	`D3.D20` and then press **Enter**

To format the monthly payments

Press	**/RFC** (for *Range*, *Format*, *Currency*), and the prompt asks for the number of decimal places
Press	**Enter** to accept the suggested two and the prompt asks for the range to format
Type	`E3.E20` and then press **Enter**

Step 7 Create a Graph. Graphs are one of 1-2-3's most powerful features. If you have a computer that can display graphs, complete this step; otherwise, skip to Step 9.

To display the Graph menu

Press	**/G** (for *Graph*)

To set the graph's X-axis

Press	**X** (for *X-axis*), and the prompt asks you to enter the X-axis range
Type	`D3.D20` and then press **Enter**

To specify a data range

Press	**A** and the prompt asks you to enter the first data range
Type	`E3.E20` and then press **Enter**

To specify a graph type

Press	**TL** (for *Type*, *Line*)

To display the graph

Press	**V** (for *View*) to display the graph

Result. The graph appears on your screen. Press **Spacebar** to remove the graph from the screen, and then press **Q** (for *Quit*) to remove the Graph menu.

TIP ON OVERLAPPING LABELS

If you created a graph using 1-2-3 Release 2.01 or earlier, the labels on the X-axis overlap. (This does not happen on Release 2.2.) To fix this, press **Spacebar** to clear the graph and return to the worksheet. Press **/GOSS** (for *Graph*, *Options*, *Scale*, *Skip*), and the prompt asks you to enter a skip factor. Type **4** and then press **Enter**. Then press Q (for *Quit*) twice to clear the menu from the screen. To display the revised graph, press **F10**. The labels on the X-axis no longer overlap. Press **Spacebar** to remove the graph.

Step 8 Change Variables. You can now change the interest rate or loan period on the template and watch the range of monthly payments change automatically. To make the change, you just type new data over old, and then press **Enter**. You can display the graph you created just by pressing **F10** and remove it by pressing **Spacebar**.

Step 9 Save Your Changes. When you make revisions to a template, you have to save them onto the disk, or they will be lost when you turn off the computer or retrieve another template.

To save your changes

Press	**Alt-M** to display the template menu
Press	**S** (for *Save*)

Result. The disk drive spins as the template is saved onto the disk. The template is designed so that when you save it protection is turned on automatically. You have to turn it off again if you want to revise your work. If you are finished with the AUTOLOAN template, press **A** (for *AUTO123*), and then press **Enter** to return to AUTO123 so that you can retrieve another template or quit the program.

EXERCISE 2
COST-VOLUME-PROFIT ANALYSIS

In this exercise, you solve cases that are based on the principles of single-product cost-volume-profit analysis discussed in Chapter 3 of *Cost Accounting: A Managerial Emphasis* (8th edition). This exercise assumes that the template CVP1 is displayed on the screen. (If you need help on retrieving files, refer to Exercise 1 or Appendix C.)

UNDERSTANDING THE TEMPLATE

The CVP1 template (Figure 9 and Table 3) has three parts into which you enter data.

- **Part 1. Breakdown of Costs** contains cells where you enter variable and fixed cost data.
- **Part 2. Cost-Volume-Profit Analysis** automatically calculates breakeven sales and other information relevant to cost-volume-profit analysis. Part 2 also allows you to input data for what-if analysis.
- **Part 3. Cost-Volume-Profit Chart Data** calculates costs and after-tax net income for a range of sales that you specify. If your computer is equipped to display graphics, you can display the data in this part on the screen as a cost-volume-profit chart or profit-volume chart.

EXPLORING THE TEMPLATE

Before you begin using the CVP1 template to solve the cases, let's use it to review the single-product cost-volume-profit example in Chapter 3 of the textbook. This allows you to concentrate on using the computer, 1-2-3, and the template since the solution to the example is already familiar.

As you saw in the text, Mary plans to sell a software package at a computer convention. She may purchase this software for $120 each and return any unsold units. The booth rental is $2,000. The units will be sold for $200 each. What do sales revenues have to be to breakeven? (Ignore income taxes, which we consider later in this exercise.)

Step 1 Enter Variable and Fixed Costs. Variable and fixed costs are entered in Part 1, where four lines are provided to list items making up each of these cost categories. The totals of any variable or fixed costs entered in this part are calculated automatically and carried to Part 2, where they are used to calculate the breakeven point and other information.

To enter variable and fixed cost labels and values

Press	**Home** (if necessary) to return to Part 1
Goto	A5, type `Unit variable cost` and then press **Enter**
Goto	G5, type `120/200` and then press **Enter**
Goto	A14, type `Fixed cost` and then press **Enter**
Goto	G14, type `2000` and then press **Enter**

FIGURE 9 The CVP1 Template.

	A	B	C	D	E	F	G
1	Part 1: Breakdown of costs						
2	============	============	============	============	============	============	============
3	Variable costs (enter as %)						Amount
4	------------	------------	------------	------------	------------	------------	------------
5	Unit variable costs						%
6							%
7							%
8							%
9	------------	------------	------------	------------	------------	------------	------------
10	Total variable costs						0.0%
11	============	============	============	============	============	============	============
12	Fixed costs (enter total amounts)						Amount
13	------------	------------	------------	------------	------------	------------	------------
14	Fixed costs					$	
15						$	
16						$	
17						$	
18	------------	------------	------------	------------	------------	------------	------------
19	Total fixed costs						$0
20	============	============	============	============	============	============	============
21	Part 2. Cost-Volume-Profit Analysis (contribution-margin method)						
22	============	============	============	============	============	============	============
23			A. ENTER WHAT IF DATA				
24	Enter after tax target net income (in $)					$	
25	Enter tax rate (in %)						%
26	Enter upper end of relevant range (in $)					$	
27	Enter additional fixed costs incurred above relevant range					$	
28							
29		B. COST-VOLUME-PROFIT CALCULATIONS					
30	Sales revenues needed to breakeven (in $)						NA
31							------------
32	Sales revenues						NA
33	Variable costs						NA
34	Contribution margin			Ratio:	NA	Dollars	NA
35	Fixed costs						NA
36	Net income before taxes						NA
37	Taxes						NA
38	Net Income after taxes						NA
39	Profit margin as a % of sales						NA
40	============	============	============	============	============	============	============
41	Part 3: Cost-Volume-Profit Chart Data						
42	============	============	============	============	============	============	============
43	Enter beginning sales revenues					$	
44	Enter sales revenues increment					$	
45	============	============	============	============	============	============	============
46	Sales	Variable	Fixed	Total		Net Income	Profit
47	revenues	costs	costs	costs	Taxes	After tax	margins
48	------------	------------	------------	------------	------------	------------	------------
49	NA	NA	NA	NA	NA	NA	NA
50	NA	NA	NA	NA	NA	NA	NA
51	NA	NA	NA	NA	NA	NA	NA
52	NA	NA	NA	NA	NA	NA	NA
53	NA	NA	NA	NA	NA	NA	NA
54	NA	NA	NA	NA	NA	NA	NA
55	NA	NA	NA	NA	NA	NA	NA
56	NA	NA	NA	NA	NA	NA	NA
57	NA	NA	NA	NA	NA	NA	NA
58	NA	NA	NA	NA	NA	NA	NA
59	NA	NA	NA	NA	NA	NA	NA
60	============	============	============	============	============	============	============

Table 3
THE CVP1 TEMPLATE

Row or Column	Description
Part 1. Breakdown of Costs	
5-8	The rows on which you enter variable costs.
14-17	The rows on which you enter fixed costs.
Part 2. Cost-Volume-Profit Analysis	
24-27	Data entry rows to explore what-ifs dealing with taxes and relevant ranges.
30	Sales revenues needed to breakeven (in $) are calculated by dividing total fixed costs (row 19) by the contribution-margin ratio (cell E34).
32	Sales revenues calculates the sales needed to earn a target net income. The results in this cell change only when you enter what-ifs in Part 2A.
33	Variable costs are calculated by multiplying the variable cost percentage (row 10) times the sales revenues (row 32).
34	The contribution-margin ratio is calculated by subtracting the variable cost percentage (row 10) from 100%. The margin in dollars is calculated by multiplying the contribution-margin ratio (cell E34) times sales revenues (row 32).
35	Fixed costs are carried from total fixed costs on Part 1 (row 19) plus additional fixed costs (row 27) if sales are above the upper end of the relevant range specified on row 26.
36	Net income before taxes is calculated by subtracting fixed costs (row 35) from the contribution margin (row 34).
37	Taxes are calculated by multiplying the tax rate you enter (row 25) times net income before taxes (row 36).
38	Net income after taxes is calculated as net income before taxes (row 36) minus taxes (row 37).
39	Profit margin is calculated as net income after taxes (row 38) divided by sales revenues (row 32).
Part 3. Cost-Volume-Profit Chart Data	
A	Sales revenues begin at the starting value on row 43 and "step" by the increment value on row 44.
B	Variable costs are calculated by multiplying the total variable cost percentage in Part 1 (row 10) times the sales revenues on the same row in column A.
C	Fixed costs are carried to this column from Part 1 (row 19).
D	Total costs are calculated by adding variable costs (column B) and fixed costs (column C).
E	Taxes are calculated by multiplying the tax rate in cell G25 times sales revenues (column A) minus total costs (column D) when sales revenues minus total costs is a positive number.
F	Net income is calculated by subtracting total costs (column D) and taxes (column E) from sales revenues (column A).
G	Profit margin is calculated by dividing the net income after taxes (column F) by sales revenues (column A). When sales are zero the program tries to divide the profit figure by 0 causing an error message (*ERR*) to be displayed in the profit margin column on this row. To prevent this from happening, the formula is nested in an @IF statement, which reads: "If sales revenues are greater than zero, divide profit after taxes by sales revenues; otherwise, display *NA*."

FIGURE 10 The CVP1 Template. After entering variable and fixed costs, Part 1 of your CVP1 template should look like this figure.

```
G14: (C0) [W11] 2000                                              READY

     A         B         C         D         E         F         G
1  Part 1: Breakdown of costs
2  ====================================================================
3  Variable costs (enter as %)                                   Amount
4  --------------------------------------------------------------------
5  Unit variable cost                                             60.0%
6                                                                     %
7                                                                     %
8                                                                     %
9  --------------------------------------------------------------------
10 Total variable costs                                           60.0%
11 ====================================================================
12 Fixed costs (enter total amounts)                             Amount
13 --------------------------------------------------------------------
14 Fixed cost                                                    $2,000
15                                                          $
16                                                          $
17                                                          $
18 --------------------------------------------------------------------
19 Total fixed costs                                             $2,000
20 ====================================================================
19-Dec-90  10:00 AM
```

Result. The totals for entries made in the variable and fixed cost sections are automatically calculated on rows 10 and 19 (Figure 10).

Variable costs are normally entered as a percentage. To make your results match those in the text, you entered this number as a fraction, 120/200. (The slash is the program's operator for division.) The result, 60%, is displayed. If you need help entering labels and numbers, see Appendix E, "Entering Data into a Template."

Step 2 Analyze Breakeven. To move to Part 2 (Figure 11), just press **PgDn**. This part has two sections. Part 2A is where you enter changes to explore what-ifs as you will see later in this exercise. Part 2B contains formulas that calculate breakeven and related values. The unit variable costs and fixed costs you entered in Part 1 have been automatically carried to Part 2B.

Step 3 Explore a Range of Possibilities. Press **PgDn** to move to Part 3 to calculate data for a wide range of sales. In this part, ranges of data are calculated for the sales revenues in column A. You can control the range of sales in column A by specifying a beginning sales figure and an increment. Entering a beginning sales revenue in cell G43 sets the sales revenues at which the range in column A begins. Entering a sales revenue increment in cell G44 establishes the increment, or steps, for the next ten rows.

Let's set the range of sales revenues so that it includes the sales revenues needed to break even that was calculated in Part 2 (cell G30). Since this was $5,000, a starting value of $0 and an increment of $1,000 gives a range of sales revenue from $0 to $10,000 in steps of $1,000 – placing the current breakeven point of $5,000 in the middle of the range.

To scale the range of sales revenues

Goto	G43, type `0` and then press **Enter**
Goto	G44, type `1000` and then press **Enter**

FIGURE 11 The CVP1 Template Calculations. Part 2 calculates breakeven and related values based on the costs that you enter into Part 1.

```
G39: (P1) [W11] 0                                                          READY

     A         B          C         D          E          F          G
21 Part 2. Cost-Volume-Profit Analysis (contribution-margin method)
22 ====================================================================
23                        A. ENTER WHAT IF DATA
24 Enter after tax target net income (in $)                       $
25 Enter tax rate (in %)                                                     %
26 Enter upper end of relevant range (in $)                       $
27 Enter additional fixed costs incurred above relevant range     $
28
29                   B. COST-VOLUME-PROFIT CALCULATIONS
30 Sales revenues needed to breakeven (in $)                            5,000
31                                                                  ----------
32 Sales revenues                                                       5,000
33 Variable costs                                                       3,000
34 Contribution margin              Ratio:    40.0%      Dollars        2,000
35 Fixed costs                                                          2,000
36 Net income before taxes                                                  0
37 Taxes                                                                    0
38 Net Income after taxes                                                   0
39 Profit margin as a % of sales                                         0.0%
40 ====================================================================
19-Dec-90  10:02 AM
```

Result. As soon as you enter the second number and then press **Enter**, Part 3 recalculates (Figure 12).

FIGURE 12 The CVP1 Chart Data. Part 3 of the template calculates ranges of data so that a CVP chart can be displayed. You control the ranges by entering beginning sales and a sales increment.

```
F44: PR [W12]                                                              READY

      A          B          C          D          E          F          G
41 Part 3: Cost-Volume-Profit Chart Data
42 ====================================================================
43 Enter beginning sales revenues                                          $0
44 Enter sales revenues increment                                       1,000
45 ====================================================================
46    Sales   Variable     Fixed      Total              Net Income    Profit
47  revenues     costs     costs      costs     Taxes     After tax   margins
48 --------------------------------------------------------------------
49         0         0     2,000      2,000         0      (2,000)        NA
50     1,000       600     2,000      2,600         0      (1,600)   -160.0%
51     2,000     1,200     2,000      3,200         0      (1,200)    -60.0%
52     3,000     1,800     2,000      3,800         0        (800)    -26.7%
53     4,000     2,400     2,000      4,400         0        (400)    -10.0%
54     5,000     3,000     2,000      5,000         0            0      0.0%
55     6,000     3,600     2,000      5,600         0          400      6.7%
56     7,000     4,200     2,000      6,200         0          800     11.4%
57     8,000     4,800     2,000      6,800         0        1,200     15.0%
58     9,000     5,400     2,000      7,400         0        1,600     17.8%
59    10,000     6,000     2,000      8,000         0        2,000     20.0%
60 ====================================================================
05-Dec-90  08:49 AM
```

Step 4 Analyze the Cost-Volume-Profit Chart. If your computer is equipped to display graphics, you can use the template menu to display charts or graphs of the data in Part 3. If not, skip ahead to Step 6.

To display the cost-volume-profit chart

Press	**Alt-M** to display the template menu
Press	**G** (for *Graphs*)
Press	**1** (for *1-CVP*)

TIP ON EXPLORING PROBLEMS ON YOUR OWN

As you explore problems on your own, you can always scale Part 3 so that the relevant range is included by just varying the beginning and increment values. You can also calculate fixed costs, variable costs, total costs, taxes, net income after taxes, and profit margin for any specific sales revenues. For example, if you want to know these values for $20,000 in sales revenue, just enter that number as the beginning sales revenue. The breakdown for that level of sales will then be shown on row 49.

Result. The cost-volume-profit chart (Figure 13) shows the relationship between costs, volume, and profit by plotting three ranges from Part 3 against the X-axis (sales revenues). The graph plots fixed expenses above variable expenses because this form is preferred by many accountants and because it emphasizes contribution-margin. Both the sales and variable-cost lines start at the origin; the vertical distance between them is the contribution-margin.

Step 5 Analyze the Profit-Volume Chart. Now, let's look at the profit-volume chart.

To display the profit-volume chart

Press **Spacebar** to remove the graph
Press **2** (for *2-P/V Chart*)

Result. The profit-volume chart (Figure 14) plots net income against sales revenues. This form is preferred by many managers who are interested mainly in the impact of changes in volume on net income. The profit-volume chart provides a comparison of how alternatives on pricing, variable costs, or fixed costs may affect net income as volume changes.

Step 6 Print the Template. You can print this template, and all others described in this book, using the Print choice on the template's main menu. This command automatically prints the template on 8½-by-11-inch paper. Before using this command, make sure your printer is on and the paper in it is aligned at the top of the page.

To print the template

Press **Spacebar** to clear the graph from the screen
Press **B** (for *Backup*) to return to the main menu
Press **P** (for *Print*)

Result. The template begins printing out on your printer. (If for any reason you want to stop the printer, hold down **Ctrl** and press **Break** repeatedly until the mode indicator reads *READY*.) When the template is finished printing, if necessary, advance the paper in the printer, and tear off the page containing your printout.

Step 7 Explore What-Ifs. Throughout the text, you almost always work with single-number "best estimates" in order to emphasize and simplify various important relationships. In short, you often conveniently assume a world of certainty.

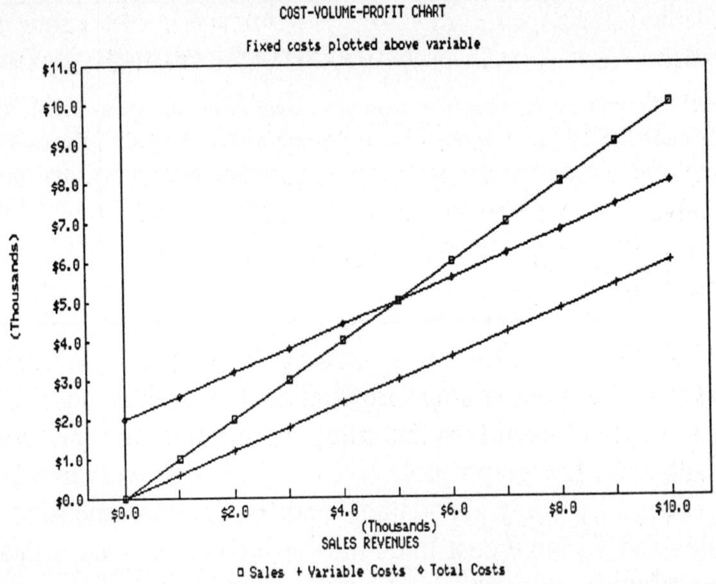

FIGURE 13 The Cost-Volume-Profit Chart. The CVP chart plots data calculated in Part 3 of the template. Here sales start at 0 and increment on each row by 1000. Sales revenues on the X-axis (from column A) range from $0 on the left to $10,000 on the right and indicate the volume of activity. Variable costs (from column B) begin at $0 and increase at the variable cost percentage on row 10. Total costs (from column D) are the sum of variable costs and fixed costs. Sales revenue increases as sales revenues on the X-axis increase. The breakeven point is where the sales revenue line meets the total cost line.

TIP ON REMOVING GRAPHS FROM THE SCREEN

On most computers, you have to press **Spacebar** to clear the graph from the screen and return to the template and template menu.

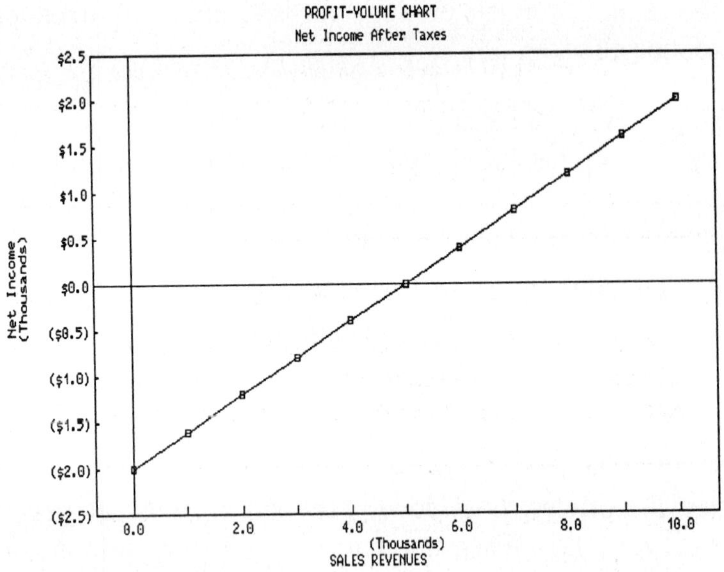

FIGURE 14 The Profit-Volume Chart. The profit-volume chart shows how profits are related to sales volume. Sales revenues on the X-axis range from $0 on the left to $10,000 on the right and indicate the volume of activity. Net income increases as sales revenues increase. It begins with a loss of $2,000 (the fixed costs) when sales revenue is zero and increases to a profit of $2,000 when sales revenue is $10,000. Breakeven is reached when the net income equals 0, in this case at $5,000 in sales.

Obviously, your estimates and predictions are subject to varying degrees of uncertainty, which is defined here as the possibility that the actual amount deviates from the expected amount. How do you cope with uncertainty? Many complex approaches are available that formally analyze expected values in conjunction with probability distributions. But the application of sensitivity analysis is the most widely used approach.

Sensitivity analysis is a what-if technique that asks how a result changes if the original estimates or projections are not achieved. In the context of cost-volume-profit analysis, sensitivity analysis answers such questions as "What is net income if volume changes?" Let's use the template to speed the process of exploring a series of what-if questions.

What-If 1 A Target Net Income Is Established. Introduce a target net income by asking, "What sales revenue must be generated to yield a target net income of $600?" If the template menu is still displayed, press **Q** (for *Quit-menu*) so that the mode indicator reads *READY*.

To enter a target net income

Goto G24, type **600** and then press **Enter**

Result. The template immediately calculates the following new results.

Row	Description
30	Sales revenues needed to break even (in $) remains $5,000.
32	Sales revenues increase to $6,500 needed to earn the target net income of $600.
33	Variable costs increase to $3,900 since the variable cost percentage is multiplied times higher sales revenues.
34	The contribution-margin ratio remains unchanged (40%), but contribution-margin dollars have increased along with the higher sales revenues.
35	Fixed costs remain unchanged at $2,000.
36	Net income before taxes is now $600, the target net income.
37	Taxes are $0 because a tax rate has not yet been entered in Part 2A.
38	Net income after taxes is $600 because there are no taxes to deduct from net income before taxes.
39	Profit margin is 9.2%.

TIP ON CALCULATING VALUES

You can use Part 3 to confirm the calculations in Part 2 of the template. Enter **6500** as the beginning sales revenue on row 43. Row 49 shows the results of all the calculations. As you would expect, the net income is $600, and the profit margin is 9.2%.

What-If 2 Income Taxes Are Considered. Suppose you are asked to compute the sales revenue needed to yield a net income after taxes of $600, the same amount calculated earlier in What-If 1. Assume an income tax rate of 30%.

To enter income taxes

Goto G25, type **.3** and then press **Enter**

Part 2. When you compare these results to the ones in the previous what-if, you see some changes. The sales revenues needed to earn the after-tax target net income rise to $7,143. The profit before taxes increases from $600 to $857 to achieve the after-tax profit of $600. Notice how the same profit after tax results in a lower profit margin (8.4%), however, because the same profit is based on a higher level of sales revenue.

Part 3. This part also shows the effects of taxes. All figures are the same as before (make sure that zero is entered on row 43 and 1000 on row 44), except the tax and net income after tax columns. There are no taxes until breakeven is reached. After-tax net income remains unchanged until after breakeven, when taxes begin to be deducted from pretax net income.

If your computer can display graphics, the profit-volume chart can help you visualize the impact of income taxes.

To display the profit-volume chart

Press	**Alt-M** to display the template menu
Press	**G** (for *Graphs*) to display the Graph menu
Press	**2** (for *2-P/V chart*)

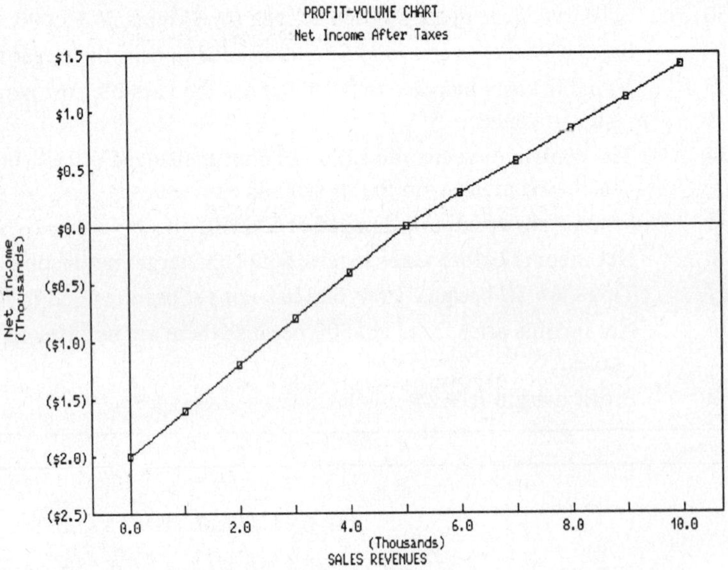

FIGURE 15 The Profit-Volume Chart. The PV chart now shows that the slope of the profit line changes because of taxes on profits after breakeven is reached.

Result. The profit-volume chart is displayed (Figure 15). Notice how the slope of the line turns slightly downward after breakeven. The slope changes because there are no income taxes up to breakeven, but in this example, income taxes add 30% to variable costs after breakeven, thereby reducing profitability and the slope of the profit-volume line.

What-If 3 Relevant Range Is Exceeded. The many assumptions that underlie the cost-volume-profit chart are subject to change if actual volume falls outside the relevant range that was the basis for drawing the chart. The relevant range is the volume range in which the firm has had some recent experience. Let's take a look at what happens if sales exceed $7,000, which requires the rental of additional space, for an increase in fixed costs of $1,000.

To remove the graph and menu

Press **Spacebar** to remove the graph
Press **B** (for *Backup*) to return to the main menu
Press **Q** for (*Quit-Menu*) to return to the worksheet

To enter the relevant range and additional fixed costs

Goto G26, type **7000** and then press **Enter**
Goto G27, type **1000** and then press **Enter**

Sales revenues (row 32) needed to cover the increased fixed costs and earn a target after-tax profit of $600 increase to $9,643. Fixed costs (row 35) increase to $3,000. Now let's look at Part 3.

Part 3 shows that fixed costs are $2,000 until the upper end of the relevant range of $7,000 is met or exceeded in the sales revenues column. From that point on, fixed costs are $3,000. Notice that you earn a profit margin of 4.7% if sales are $6,000, but you have a loss of 2.9% if you sell 7,000 units. In fact, the profit margin doesn't return to 4.7% until sales are $9,000. This is because you incur the added fixed costs of $1,000 when volume exceeds the relevant range. Another look at Part 3 indicates that we now have two breakeven points: the old breakeven of $5,000 and a new breakeven outside the relevant range between $7,000 and $8,000. This point becomes clearer by examining the graphs (Figures 16 and 17).

FIGURE 16 The Cost-Volume-Profit Chart. The cost-volume-profit chart shows how total costs reflect the jump in fixed costs. Breakeven occurs twice — at a sales level of $5,000 ($5.0) and once again at sales above $7,000 ($7.0).

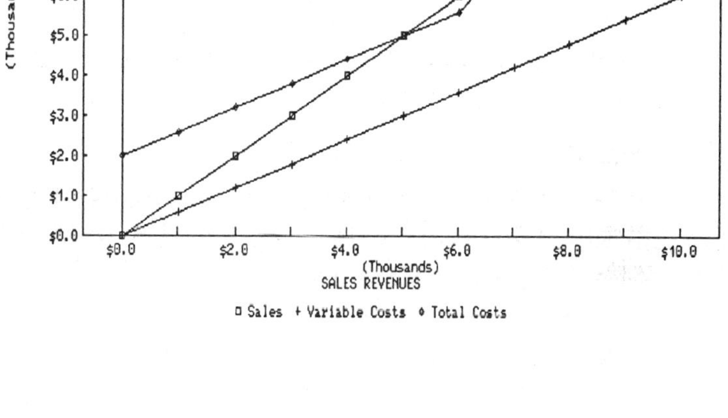

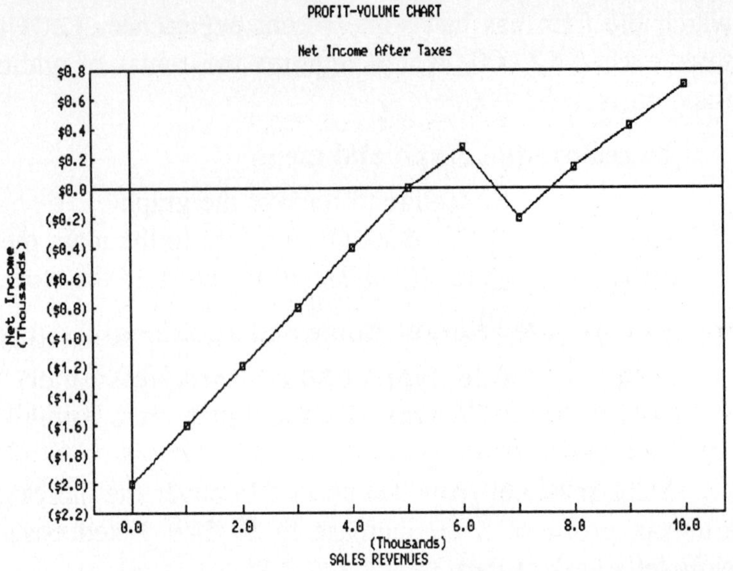

FIGURE 17 The Profit-Volume Chart. The profit-volume chart shows that profits reach about $300 at sales of about $6,000 but then plummet to about -$200 when sales exceed the upper end of the relevant range. It takes sales of over $8,000 to get profits back to their original level.

TIP ON INTERPRETING THESE CHARTS

1-2-3 plots "data points," which are actual values, and then connects these data points with straight lines. In this case, when sales reach $7,000, fixed costs increase by $1,000, as can easily be seen on Part 3. Since Part 3 is calculating data points in increments of $1,000, the charts show misleading information about what's actually happening between $6,000 and $7,000 – you can't interpolate between these data points. To get a better idea of what is happening, change the beginning sales on Part 3 to $6,000 and the increment to $200 – to look at this critical range.

Step 8 Save Your Work. If you are in the middle of an analysis and want to quit, you have to save the template along with the numbers you have already entered if you want to be able to return to them later. Since each template has a command on its own template menu that clears data from the template, feel free to save your work when you want to quit for the day. Any time you want, you can clear the data with this command and return the template to its blank state. Be sure you have the *Cost Accounting Resource Disk* in the drive from which you retrieved the file.

To save the template

Press **Alt-M** to display the template menu
Press S (for *Save*)

Result. A message appears on the screen, the disk drive spins, and its light comes on. In a few seconds, the light goes out, and the highlight on the screen jumps to *Press Enter to continue*. Press **Enter** to display the template and the template menu.

Step 9 Clear Data from the Template. In the sections that follow, you explore cases; therefore, the data you have entered to get acquainted with the template have to be removed. The template menu Clear command does this for you automatically. This command restores the template to its original state, and you can use it at any time.

When clearing other templates, if you want to save your data, be sure you save the template before using the command.

To clear data from the template

Press **C** (for *Clear*)

Result. When you select Clear, a frame appears on the screen warning you that this command erases your data from the disk. If you want to prevent this, hold down **Ctrl** and press **Break**. This cancels the command and clears the menu from the screen. Press **Home** to return to the template. If you want to erase the data, just press **Enter**.

Note: The next time you retrieve the template it contains the data you saved in Step 8 but just erased in this step. If you want to save the template in its original state, select the Save choice from the menu to save the blank template and overwrite the previous version currently on the disk.

CASE 1 FRED WILLIAMS DRY CLEANING

Fred Williams owns a chain of fine dry cleaning stores located strategically to serve the population of a medium-sized city. His chain is successful because of excellent customer service as well as competitive prices. He advertises that "You live within 5 minutes of a Williams Cleaners" and "Our friendly people believe that customer service comes first."

The chain contracts with another company to do the dry cleaning at a cost averaging 45% of sales dollars. Other store operating costs are primarily fixed, and the marginal income tax rate is 35%. The most recent operating statement appears below.

Fred Williams Dry Cleaning
INCOME STATEMENT
For the Year Ended 12/31/x1

Sales		$900,000
Cost of Sales (45% of Sales)		405,000
Gross Margin		$495,000
Operating Expenses (Fixed)		
Store Rentals	$ 60,000	
Store Utilities	28,000	
Store Personnel – Salaries & Benefits	300,000	$388,000
Net Income Before Tax		$107,000
Provision for Income Taxes (35%)		37,450
Net Income		$ 69,550

Fred is becoming concerned that his stores may not be providing enough service in the form of convenient hours of operation. He notices that other retail stores in the city are open more hours to accommodate working people. His sales have been declining over the last few months, and Fred is convinced that this is because other dry cleaning stores in the area are open more hours than his stores.

To expand his hours of operation, Fred could hire some part-time employees. However, he is afraid that they would not provide the quality of service his chain is known for. After talking with his current employees, he finds that they are agreeable to staggered work hours. This allows Fred to hire one more full-time employee per store, which would result in the stores being open an additional 20 hours per week. After working out this work schedule, which we term Plan A, Fred realizes that he could hire a second full-time employee for each store. This plan would add 15 more hours of service per week. Store hours would not increase as much with this plan because now the stores would be open the maximum hours per week that appear to have business potential. This latter plan we call Plan B.

The most likely estimates for these two plans as compared to the current situation are as follows:

	Plan A	Plan B
Increased Salary and Benefits Cost	$ 50,000	$100,000
Increased Sales	100,000	200,000

The salary and benefits cost estimate is a certainty because it reflects the market salary in the city, and the unemployment rate is such that there are always applicants available at this salary. The sales increase estimate, however, is mostly based on regaining the business lost when the competition began longer hours of service. Consequently, it is subject to some uncertainty.

REQUIREMENTS

A. Determine the breakeven sales and the minimum increase in sales required so that net income does not decline for both Plan A and for Plan B. Use the "additional fixed costs above the relevant range" feature in Part 2A of the template. (*Hint:* In Part 2A of the template, enter a target net income of $69,550.)

B. Compute the net income for both Plan A and Plan B using their projected sales increases. (*Hint:* In Part 3 of the template, enter a beginning sales of $900,000, and then select a sales increment to calculate net income at the increased sales estimate for the plan being analyzed.)

C. Prepare a recommendation for Fred Williams. You may consider and report sales levels other than those already used in Requirements A and B. (*Hint:* Review Part 3 of the template. You need only one printout for each plan to respond to all the requirements if you take advantage of the hints.)

Name:________________________

CASE 1. FRED WILLIAMS DRY CLEANING SOLUTION
Template: CVP1

DATA ANALYSIS SUMMARY (attach model printouts as supporting documents):

Fred Williams Dry Cleaning Company
Analysis of Increased Operating Hours Plans
Minimum Sales Estimate

REQUIREMENT A	BREAKEVEN SALES	MINIMUM SALES ESTIMATE	AFTER-TAX NET INCOME	PROFIT MARGIN
Current situation	$705,455	$900,000	$69,550	7.7%
Plan A				
Plan B				

Fred Williams Dry Cleaning Company
Analysis of Increased Operating Hours Plans
Most Likely Sales Estimate

REQUIREMENT B	MOST LIKELY SALES ESTIMATE	AFTER-TAX NET INCOME	PROFIT MARGIN
Plan A			
Plan B			

DISCUSSION

CASE 2 MARY ANN COSMETICS

You are the sales manager for the western region of the Mary Ann Cosmetics Company. Because your bonus is based on the net income of your region, you are highly motivated to have a very profitable region.

Mary Ann Cosmetics produces and distributes its products to the regions and provides all the national advertising. As a regional manager, you buy the product from the parent company at 45% of the retail sales price. Costs of distributing the product to the sales staff average 10% of sales. Since your sales force is all in the field (they basically operate out of the trunks of their cars) and the parent company does all the advertising, the fixed costs of operating your region are only $100,000 per year.

The only other expenditure you must make is to pay your sales force. Currently, each member of the sales force has an exclusive territory and is on salary. The total salaries for the sales force for the year are $800,000. Regional performance for last year is presented below.

Mary Ann Cosmetics Company
Western Region – Income Statement
For the Year Ended 12/31/x1

Sales		$3,000,000
Cost of Sales (45% of Sales)		1,350,000
Gross Margin		$1,650,000
Operating Expenses		
Sales Salaries (Fixed)	$800,000	
Distribution Costs (10%)	300,000	
Other Operating (Fixed)	100,000	1,200,000
Net Income		$ 450,000

Knowing the bonus plan motivates you to have a profitable region, you begin to wonder if you can stimulate the sales force in a like manner. You begin to consider alternative compensation plans hoping to stimulate more sales and, therefore, greater bonuses for yourself.

Industry sources indicate that a straight commission plan (no salary payments) with a 28% commission on sales could stimulate a $300,000 increase in sales above the existing level. Alternatively, cutting salaries by 50% and substituting a 15% commission on sales could stimulate a $250,000 increase in sales. A third approach is to pay a profit sharing bonus of 2% of regional net income and keep the salaries as they are. This alternative would stimulate a $100,000 sales increase.

Assume the net income of your region is computed on a pretax basis.

TUTORIAL 2 ADDING A NEW PART TO THE CVP1 TEMPLATE

The CVP1 template has room to enter four items that make up variable costs. Let's see how you add a section to expand the number of lines where these costs can be itemized. When you have completed this tutorial, your template will look like Figure 18.

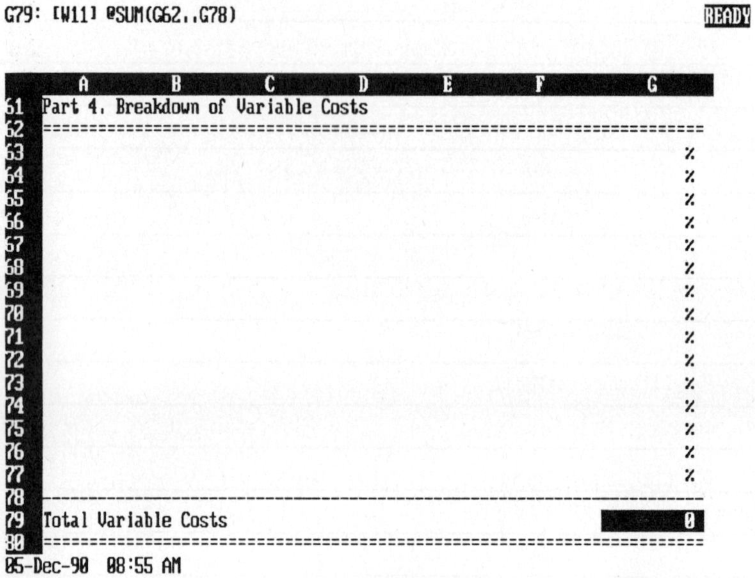

FIGURE 18 The Revised CVP1 Template. After entering a new Part 4 to the CVP1 template, your template will look like this figure.

Step 1 Retrieve the CVP1 Template. Retrieve the template if it isn't already on the screen.

Step 2 Turn Off Protection. Turn off protection so that you can enter data into cells on the worksheet.

To turn off protection

Press **/WGPD** (for *Worksheet, Global, Protection, Disable*)

Step 3 Enter Label. This new section will be Part 4 of the template.

To enter a heading for the new section

Goto A61, type `Part 4. Breakdown of Variable Costs` and then press **Enter**

Step 4 Enter Double-Ruled Lines. Doubled-ruled lines, made up of equal signs, can be entered using 1-2-3's repeating label command. This is the backslash key (\) in the lower left corner of the keyboard. When you press this key, the next character or characters that you type will be repeated across the width of the cell. This cell-wide line can then be copied elsewhere on the worksheet.

To enter a double-ruled line in a cell

Goto A62, press \= and then press **Enter**

To copy the double-ruled line to other columns

Goto	A62, press **/C** (for *Copy*), and the prompt asks for the range to copy FROM
Press	**Enter**, and the prompt asks for the range to copy TO
Type	**B62.G62** and then press **Enter**

To copy the double-ruled line to another row

Press	**/C** (for *Copy*), and the prompt asks for the range to copy FROM
Type	**A62.G62** and then press **Enter**. The prompt asks for the range to copy TO
Type	**A80** and then press **Enter**

Result. The doubled-ruled line is copied so that it fills rows 62 and 80.

Step 5 Enter a Single-Ruled Line. Single-ruled lines, made up of minus signs, can be entered the same way as double-ruled lines.

To enter a single-ruled line in a cell

Goto	A78, press \- and then press **Enter**

To copy the single-ruled line to other columns

Goto	A78, press **/C** (for *Copy*), and the prompt asks for the range to copy FROM
Press	**Enter**, and the prompt asks for the range to copy TO
Type	**B78.G78** and then press **Enter**

Result. The single-ruled line now fills row 78.

Step 6 Copy Data Entry Spaces from Part 1. Part 1 has blank data entry spaces that can be quickly copied into this new section.

To copy data entry spaces from Part 1

Press	**/C** (for *Copy*), and the prompt asks for the range to copy FROM
Type	**G5** and then press **Enter**. The prompt asks for the range to copy TO
Type	**G63.G77** and then press **Enter**

Result. The data entry spaces are copied from Part 1. If you had data on row 8, that data would be copied instead of the data entry spaces. Use the Clear command on the template menu, and then repeat the procedure.

Step 7 Enter a Function to Sum the Rows. Functions are small programs that are built into 1-2-3. They frequently perform complicated calculations and always save you a great deal of time. Here you enter the function that sums the values entered in a range.

To enter a formula that sums data entered in column G

Goto	A79, type **Total Variable Costs** and then press **Enter**
Goto	G79, type **@SUM(G62.G78)** and then press **Enter**

To format the cell with the formula

Press	**/RFP** (for *Range*, *Format*, *Percent*), and the prompt asks for the number of decimal places
Type	**1** and then press **Enter**. The prompt asks for the range to format
Press	**Enter**

Step 8 Link the New Section to Part 1. Now that Part 4 is finished, the final step is to link it to Part 1 so that any numbers entered in this section are added to the numbers totaled from Part 1. To do this, you edit the existing function in cell G10.

To link the new section with Part 1

Goto	G10, and then press **Edit (F2** on the IBM PC) to display the formula *@SUM(G4..G9)* on the Edit line
Type	**+G79** so that the formula reads *@SUM(G4..G9)+G79*
Press	**Enter**

Result. You can now enter data in Part 4, and it is carried to Part 1 automatically. Enter .25 in any cell in column G of Part 4, and then press **Enter**. Return to Part 1, and you see it has been carried to row 10.

Step 9 Save Your Changes. Now that you have revised your template you should save it.

To save your changes

Press	**Alt-M** to display the template menu
Press	**S** (for *Save*)

Result. The revised template is saved onto the disk, overwriting the previous version. The template menu's Save command is designed so that it turns on protection whenever you use it. If you want to revise your template further, you have to turn protection off again.

When using the new section, keep it mind that it isn't cleared automatically by the Clear command on the template menu. To erase data you enter in the new section, move the cell pointer to Part 4, and then press **/RE** (for *Range*, *Erase*). When asked to enter the range to erase, type **A63.G77** and then press **Enter**.

EXERCISE 3
MULTIPLE-PRODUCT COST-VOLUME-PROFIT ANALYSIS

In this exercise, you solve cases that are based on the principles of multiple product cost-volume-profit analysis discussed in Chapter 3 of *Cost Accounting: A Managerial Emphasis* (8th edition). This exercise assumes that the template CVP2 is displayed on the screen. (If you need help on retrieving files, refer to Exercise 1 or Appendix C.)

UNDERSTANDING THE TEMPLATE

The CVP2 template has two parts (Figure 19 and Table 4).

- **Part 1. Multiproduct Company** is where you enter data for up to four products. Formulas in this part then calculate breakeven sales, contribution margin ratios, and other information relevant to multiple-product cost-volume-profit analysis.
- **Part 2. CVP Chart Data for Total Column on Part 1** calculates costs and net income for a range of sales that you specify. The data in this part can also be displayed on the screen as a cost-volume-profit chart or profit-volume chart if your computer is equipped to display graphics.

EXPLORING THE TEMPLATE

Before you begin using the CVP2 template to solve cases, let's use it to review the multiple-product cost-volume-profit example in Chapter 3 of the textbook. This allows you to concentrate on using the computer, 1-2-3, and the template since the solution to the example is already familiar. The example uses a two-product company with the following budget:

	Peerless Pens	Fantastic Pens	Total
Sales in Units	120,000	40,000	160,000
Sales @ $5 and $10	$600,000	$400,000	$1,000,000
Variable Costs @ $4 and $3	480,000	120,000	600,000
Contribution Margins @ $1 and $7	$120,000	$280,000	$400,000
Fixed Costs			300,000
Net Income			$100,000

Step 1 Enter Data. To begin, enter the data shown above into Part 1 of the template.

To enter data for Peerless Pens

Goto	D4, type `120000` and then press **Enter**
Goto	D5, type `5` and then press **Enter**
Goto	D6, type `4` and then press **Enter**

FIGURE 19 The CVP2 Template.

```
     A     B          C          D          E          F          G          H
 1 |Part 1: Multi-product Company
 2 |= =========== =========== =========== =========== =========== =========== ===========
 3 |A. PRODUCT BREAKDOWN             1            2            3            4         Total
 4 |Sales in units.....:     #_________ #_________ #_________ #_________               NA
 5 |Selling prices ....:     $_________ $_________ $_________ $_________               NA
 6 |Unit variable costs:     $_________ $_________ $_________ $_________               NA
 7 |Contrib. margin % .:             NA          NA          NA          NA           NA
 8 |% of total units ..:             NA          NA          NA          NA           NA
 9 |% of total sales ..:             NA          NA          NA          NA           NA
10 |
11 |B. MULTI-PRODUCT ANALYSIS
12 |
13 |Sales revenues.....:             0           0           0           0            0
14 |Variable costs.....:             0           0           0           0            0
15 |Contribution margin:             0           0           0           0            0
16 |Fixed costs.......................................................:  $_________
17 |Net income.........:                                                             NA
18 |B/E sales in units.:             NA          NA          NA          NA           NA
19 |B/E sales in $ ....:             NA          NA          NA          NA           NA
20 |= =========== =========== =========== =========== =========== =========== ===========
21 |Part 2. CVP Chart data for total column on Part 1.
22 |= =========== =========== =========== =========== =========== =========== ===========
23 |Enter beginning units for CVP chart X-axis ..........................:  #_________
24 |Enter unit sales increment for CVP chart X-axis .....................:  #_________
25 |= =========== =========== =========== =========== =========== =========== ===========
26 |      Units       Sales    Variable       Fixed       Total         Net      Profit
27 |       sold    revenues       costs       costs       costs      income      Margin
28 |= =========== =========== =========== =========== =========== =========== ===========
29 |         NA          NA          NA          NA          NA          NA          NA
30 |         NA          NA          NA          NA          NA          NA          NA
31 |         NA          NA          NA          NA          NA          NA          NA
32 |         NA          NA          NA          NA          NA          NA          NA
33 |         NA          NA          NA          NA          NA          NA          NA
34 |         NA          NA          NA          NA          NA          NA          NA
35 |         NA          NA          NA          NA          NA          NA          NA
36 |         NA          NA          NA          NA          NA          NA          NA
37 |         NA          NA          NA          NA          NA          NA          NA
38 |         NA          NA          NA          NA          NA          NA          NA
39 |         NA          NA          NA          NA          NA          NA          NA
40 |= =========== =========== =========== =========== =========== =========== ===========
```

Table 4
THE CVP2 TEMPLATE

Row or Column	Description
Part 1. Multiproduct Company	
4-6	Data entry spaces where you enter data for up to four products.
7	Formulas calculate contribution-margin ratios.
8-9	The percentage of each product relative to total units and total sales dollars.
13-19	Formulas calculate a variety of cost-volume-profit-related values based on the data entered in Part A.
16	A data entry space where you enter total fixed costs.
Part 2. CVP Chart Data for Total Column on Part 1	
B	Units sold is set by the beginning units (row 23) and unit sales increment (row 24). It assumes a constant sales mix throughout the range.
C	Sales revenues are units sold (column B) times weighted average selling price (cell H5). They assume a constant sales mix throughout the range.
D	Variable costs assume a constant sales mix throughout the range and are calculated by multiplying the weighted average unit variable cost (cell H6) times the units sold (column B).
E	Fixed costs are carried from row 16 and are constant throughout the range.
F	Total costs are calculated by adding variable costs (column D) and fixed costs (column E).
G	Net income is calculated by subtracting total costs (column F) from sales revenues (column C).
H	Profit margin is calculated by dividing the net income (column G) by sales revenues (column C).

To enter data for Fantastic Pens

Goto E4, type **40000** and then press **Enter**
Goto E5, type **10** and then press **Enter**
Goto E6, type **3** and then press **Enter**

To enter fixed expenses

Goto H16, type **300000** and then press **Enter**

Result. The template (Figure 20) shows the breakeven point for the company as a whole and the mix of products that make up this total. The company as a whole breaks even with sales of 120,000 units (row 18) and $750,000 in sales revenues (row 19). But remember, this calculation assumes that the sales mix remains the same, so 90,000 of the units required to break even must be Peerless Pen sales (75% x 120,000 units) and 30,000 must be Fantastic Pen sales (25% x 120,000). This is shown in the columns for those products.

Step 2 Explore a Range of Possibilities. Press **PgDn** to move to Part 2 of the template to calculate data for a wide range of unit sales.

To scale the graph data

Goto H23, type **0** and then press **Enter**

Goto H24, type **20000** and then press **Enter**

Result. As soon as you enter the second number and press **Enter**, the entire section recalculates (Figure 21).

FIGURE 20 The CVP2 Template. The CVP2 template automatically calculates breakeven for the company as a whole.

H16: (,0) PR 300000 READY

	A B C	D	E	F	G	H
1	Part 1: Multi-product Company					
2	============					
3	A. PRODUCT BREAKDOWN	1	2	3	4	Total
4	Sales in units.....:	120,000	40,000			160,000
5	Selling prices ....:	$5.00	$10.00			$6.25
6	Unit variable costs:	$4.00	$3.00			$3.75
7	Contrib. margin % .:	20.00%	70.00%	NA	NA	40.00%
8	% of total units ..:	75.00%	25.00%	NA	NA	100.00%
9	% of total sales ..:	60.00%	40.00%	NA	NA	100.00%
10						
11	B. MULTI-PRODUCT ANALYSIS					
12						
13	Sales revenues.....:	600,000	400,000	0	0	1,000,000
14	Variable costs.....:	480,000	120,000	0	0	600,000
15	Contribution margin:	120,000	280,000	0	0	400,000
16	Fixed costs.......................................:					300,000
17	Net income........:					100,000
18	B/E sales in units.:	90,000	30,000	NA	NA	120,000
19	B/E sales in $ ....:	450,000	300,000	NA	NA	750,000
20	============					

19-Dec-90 10:03 AM

FIGURE 21 CVP2's Chart Data. The chart data in Part 2 is calculated for the company as a whole.

H24: PR 20000 READY

	A B	C	D	E	F	G	H
21	Part 2. CVP Chart data for total column on Part 1.						
22	============						
23	Enter beginning units for CVP chart X-axis ....:						0
24	Enter unit sales increment for CVP chart X-axis ....:						20,000
25	============						
26	Units	Sales	Variable	Fixed	Total	Net	Profit
27	sold	revenues	costs	costs	costs	income	Margin
28	============						
29	0	0	0	300,000	300,000	(300,000)	NA
30	20,000	125,000	75,000	300,000	375,000	(250,000)	-200.0%
31	40,000	250,000	150,000	300,000	450,000	(200,000)	-80.0%
32	60,000	375,000	225,000	300,000	525,000	(150,000)	-40.0%
33	80,000	500,000	300,000	300,000	600,000	(100,000)	-20.0%
34	100,000	625,000	375,000	300,000	675,000	(50,000)	-8.0%
35	120,000	750,000	450,000	300,000	750,000	0	0.0%
36	140,000	875,000	525,000	300,000	825,000	50,000	5.7%
37	160,000	1,000,000	600,000	300,000	900,000	100,000	10.0%
38	180,000	1,125,000	675,000	300,000	975,000	150,000	13.3%
39	200,000	1,250,000	750,000	300,000	1,050,000	200,000	16.0%
40	============						

19-Dec-90 10:03 AM

Step 3 Display the CVP Chart. The cost-volume-profit chart plots sales revenues (column C), variable costs (column D), and total costs (column F) against units sold (column B).

To display the cost-volume-profit chart

Press **Alt-M** to display the template menu
Press **G** (for *Graphs*)
Press **1** (for *1-CVP*)

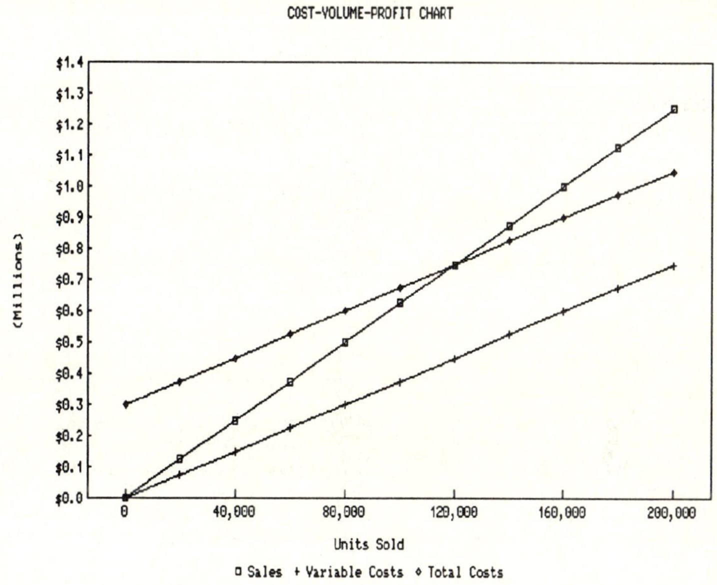

FIGURE 22 The CVP Chart. The CVP chart plots data from Part 3 against units sold.

Result. The CVP chart is displayed (Figure 22). Units sold on the X-axis (column B) indicate the volume of activity. Sales revenue (column C) increases as increased units are sold. Variable costs (column D) begin at 0 and increase as unit sales increase. Total costs (column F), plotted above variable costs, start at $300,000 (the fixed costs) and increase to over $1,000,000 as variable costs increase. The breakeven point is reached at 120,000 units, where the sales revenues line meets the total cost line.

Step 4 Display the Profit-Volume Chart. The profit-volume chart plots profits (column G) against the units sold (column B).

To display the profit-volume chart

Press **Spacebar** to clear the graph from the screen
Press **2** (for *2-P/V chart*)

Result. The profit-volume chart is displayed (Figure 23). Units sold on the X-axis (column B) indicate the volume of activity. Profits increase as increased units are sold. The breakeven point is reached at 120,000 units, where the profit line crosses the zero line of the chart.

Step 5 Clear Data from the Template. In the sections that follow, you enter new case data, so the data you have entered into the template to review Chapter 3 have to be removed. If you want to save your data, be sure you have saved the template onto a disk before using the command.

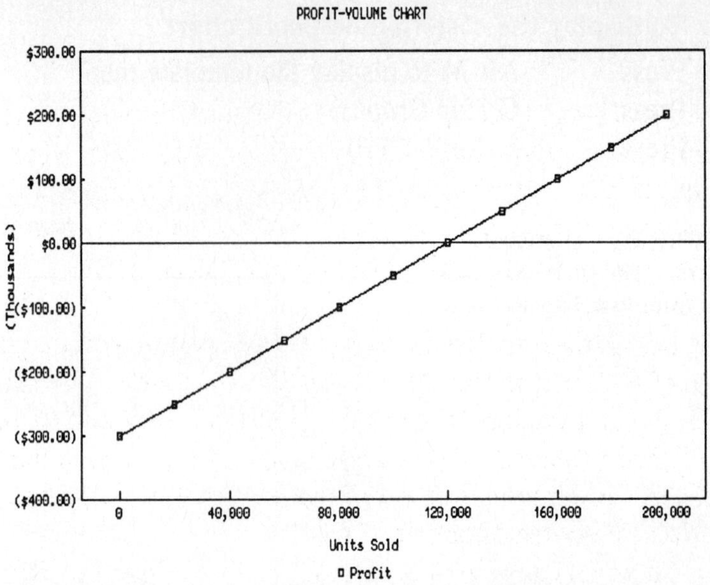

FIGURE 23 The Profit-Volume Chart. The profit-volume chart plots profits against units sold.

To clear data from the template

Press	**Spacebar** to clear the graph from the screen
Press	**B** (for *Backup*) to return to the main menu
Press	**C** (for *Clear*)

Result. When you select Clear, a frame appears on the screen warning you that this command will erase your data from the template. If you want to erase the data, just press **Enter**.

CASE 1 LEWIS & CLARK

Lewis & Clark, P.C., provides a variety of accounting services to its clients. Located in a large city, they are finding that competition from other accounting firms is becoming more intense. Through increased staff training and the acquisition of several microcomputers, Lewis & Clark has greatly increased the efficiency with which they provide their services. Despite this increased efficiency, competition is still decreasing profitability.

Bob Lewis, the senior partner, requests an analysis of the prior year's revenues to determine the profitability of each service group. He believes it is time for the firm to redirect its attention to the most profitable services to bring back the firm's faltering profitability. Based on his knowledge of the accounting profession, Bob realizes that audit fees are rapidly declining due to competition. But he needs the specific facts before he charts a course of action.

The first step in the analysis is to divide the firm's services into four primary categories — auditing, tax, consulting, and compilation and review. Next, you gather information on the hours billed, average billing rates per hour, and average direct cost per hour for each category from last year's data. This yields the following:

Category	Hours	Average Billing Rate/Hour	Average Cost/Hour
Auditing	31,200	$56.50	$29.25
Tax	15,340	74.25	31.50
Consulting	6,485	91.75	36.75
Compilation & review	8,860	52.25	29.30

As Bob suspected, the average billing rate is low for audit services. This is because of the increased number of fixed fee audits (clients are billed an amount agreed to rather than the number of hours the audit takes times the normal billing rate) that clients are demanding. Clients are more willing to pay full rates for the time provided for tax and consulting work because the benefits to the client are more evident.

All accounting firms would like to provide only the most profitable services — tax and consulting. However, they realize that clients demand a full range of services. These complimentary effects require the firm to provide a certain minimum level of other services. Further, increased staff training and the addition of people with the necessary expertise are needed to move the firm's service mix into these more profitable areas. Last year, Lewis & Clark incurred fixed costs of operating the firm (not including the four partners' draw) of $1,680,000.

REQUIREMENTS

A. Compute the breakeven revenue and the net income of Lewis & Clark for last year using their actual product mix.

B. If the firm could have shifted 3,000 hours from auditing to tax last year, how would their breakeven revenue and profitability have changed?

C. Assume that the hours of auditing and compilation and review services cannot be changed.

   C1. How many hours of tax service must be added to increase net income to $600,000? (*Hint:* Insert an increasing number of hours of tax service until net income equals $600,000.)

   C2. How many hours of consulting service must be added to increase net income to $600,000?

D. Would your answer to Requirements C1 and C2 (the additional hours of tax and consulting service, respectively, that are needed to increase net income to $600,000) differ if:

   D1. Adding tax services would require $20,000 in additional fixed costs while,

   D2. Adding consulting services would require $40,000 in additional fixed costs?

E. On the basis of these computations, what course of action would you recommend to Bob Lewis?

Name: ______________________

CASE 1. LEWIS & CLARK SOLUTION
Template: CVP2

DATA ANALYSIS SUMMARY (attach model printouts as supporting documents):

Lewis & Clark, P.C.
Product Mix Analysis – Alternatives

REQUIREMENT	CONTRIBUTION MARGIN RATIO	BREAKEVEN REVENUE	REVENUE	PROFIT
A				
B				
C1				
C2				
D1				
D2				

DISCUSSION

CASE 2 LINCOLN ELECTRIC

As a manufacturer of quality electric motors for industrial use, Lincoln Electric has a loyal group of customers. Price had not been an issue during this long-standing relationship because Lincoln Electric has a reputation for building very reliable motors that matched its customers' needs.

Lincoln's controller, Ann Archer, began budget preparation for the next fiscal year by discussing potential demand for each of the four models Lincoln produced with its key customers. She then worked with the plant manager to estimate production costs for this potential demand. Further meetings with the vice president of marketing and the company president on pricing policy, net income goals, and selling costs determined the following estimates:

MODEL	UNITS	SALES PRICE	VARIABLE COSTS
160	3,425	$ 500	$ 300
260	2,150	800	440
360	1,350	1,600	800
460	925	3,500	1,575

Total fixed costs for manufacturing, selling, and administration were predicted to be $3,265,000.

During the year, costs were incurred as projected. Because of the good customer response, no price cutting took place. Total units sold for the year exceeded the budget, but net income was almost 10% below budget. To explain this situation, the controller found that the units sold of each model were Model 160 – 3,615; Model 260 – 2,252; Model 360 – 1,233; and Model 460 – 882.

REQUIREMENTS

A. Using the budgeted data, compute the breakeven sales, projected sales, and projected net income of Lincoln Electric.

B. Using the actual product mix achieved, compute the breakeven sales, actual sales, and actual net income of Lincoln Electric. Explain to the company president why the actual net income is less than the projected net income.

C. Summarize the impact on net income of actual sales volume differing from budgeted volume for each model. (*Hint:* Start with the budgeted data for all four products; then substitute the actual data for one product at a time to observe its impact.) Explain how this refines the analysis you did in Requirement B. What would you recommend to management?

Name:__________________________

CASE 2. LINCOLN ELECTRIC SOLUTION

Template: CVP2

DATA ANALYSIS SUMMARY (attach model printouts as supporting documents):

Lincoln Electric Company
Product Mix Analysis – Actual vs. Budget

REQUIREMENT	CONTRIBUTION-MARGIN RATIO	BREAKEVEN REVENUE	SALES	NET INCOME
A				
B				

Lincoln Electric Company
Analysis of Individual Model Sales

REQUIREMENT C	CONTRIBUTION-MARGIN RATIO	NET INCOME
Model 160 – actual		
Model 260 – actual		
Model 360 – actual		
Model 460 – actual		

DISCUSSION

CASE 3 OVERNIGHT TRUCK LINES

Deregulation in the trucking industry has resulted in a flood of new companies into the industry, which has forced the industry leaders to adopt new strategies to survive. Competition in the truckload market (shipments of 10,000 pounds or more) is particularly fierce because of the ease of entry into this market and low capital investment required. Adding to the woes of the industry is the revitalization of railroads, which are aggressively marketing their piggyback service (highway trailers moved via rail). The industry leaders still hold a major share of the Less-Than Truckload (LTT) market (shipments less than 10,000 pounds) because this market requires larger capital investment in terminals, which restricts entry.

Overnight Truck Lines, Inc., a Class I motor carrier (revenues of at least $5 million), is attempting to develop new strategies in this environment but is still experiencing declining profitability. An analysis of their prior year's performance yielded the following:

SERVICE	TON-MILES (000)	REVENUE/TON-MILE (000)	VARIABLE COST/TON-MILE (000)
Truckload	679	$7.00	$3.25
LTT	857	27.75	6.30
Piggyback	394	13.30	5.05

In addition, fixed operating costs in thousands were $22,917.

Although Overnight Truck Lines, Inc., would like to deemphasize the highly competitive truckload market, it is constrained somewhat from doing so. Many of its customers in the LTT market expect that Overnight will also service their truckload needs at competitive rates. It is estimated that any increase or decrease in ton-miles for the truckload market causes a corresponding increase or decrease in the LTT market.

A further consideration in planning the service mix is that piggyback shipments have increased 50% in the last three years. If Overnight is to grow at all, it has to achieve a share of this market even if it is less profitable.

REQUIREMENTS

A. Compute the contribution-margin ratio, sales revenue, net income, and breakeven sales for the prior year.

B. Compute the contribution-margin ratio, sales revenue, net income, and breakeven sales if truckload service is reduced 25 thousand ton-miles (remember that LTT service is also reduced) while Piggyback service is increased by 50 thousand ton-miles.

C. Compute the contribution-margin ratio, sales revenue, net income, and breakeven sales if the LTT service is increased 25 thousand ton-miles for an added fixed cost, in thousands, of $400 for extra salaries and truck leasing. Remember that the truckload service should also increase by 25 thousand ton-miles.

D. What recommendations would you make to management as a result of your analysis?

Name:____________________________

CASE 3. OVERNIGHT TRUCK LINES SOLUTION
Template: CVP2

DATA ANALYSIS SUMMARY (attach model printouts as supporting documents):

Overnight Truck Lines, Inc.
Product Mix Options
(000 Omitted)

REQUIREMENT	CONTRIBUTION-MARGIN RATIO	SALES REVENUE	NET INCOME	BREAKEVEN SALES
A				
B				
C				

DISCUSSION

TUTORIAL 3 PRINTING GRAPHS

Many of the templates in this book have graphs that you can display if your computer is equipped to display graphics. Let's see how you print a graph. Even if you can't see graphs on the screen, you can print them out. Printing graphs can take quite a long time, so don't begin this tutorial unless you have at least 30 minutes to complete it. If you are using the Lotus 1-2-3 program, you will need a copy of the Lotus PrintGraph disk to complete this tutorial.

Begin by retrieving the CVP2 template if it isn't already on the screen. Enter the same data you entered in Steps 1 and 2, beginning on page 45.

Step 1 Save the Graph. Begin by displaying the graph using the program's command menu (not the template menu). The Graph Save command saves the *current* graph, which is the one you viewed most recently. If your computer isn't equipped to display graphs, you won't see the graph, but the command works nonetheless.

To display the graph

Press	**/GNU** (for *Graph*, *Name*, *Use*), and the prompt asks for the name of the graph to make current
Press	→ to highlight *GRAPH 2,* and then press **Enter** to display the graph
Press	**Spacebar** to remove the graph from the screen and return to the Graph menu

To save the graph

Press	**S** (for *Save*), and the prompt asks for the name of the file in which you want to save the graph
Type	**PV2** and then press **Enter** to save the graph on the disk
Press	**Q** (for *Quit*) to leave the Graph menu

Step 2 Quit the Lotus 1-2-3 Program and Load the PrintGraph Program. You print graphs with the Lotus PrintGraph program.

To quit 1-2-3 and load PrintGraph

Press	**/QY** (for *Quit*, *Yes*) to return to the Lotus Access System menu
Press	**P** (for *PrintGraph*)

Result. The prompt reads *Insert Lotus PrintGraph disk in drive A and press [Enter].* Insert the disk, and then press **Enter**. The drive spins, and in a moment, the PrintGraph menu appears on the screen.

Step 3 Select and Preview the Graph. You select a graph from a list of the graphs on the disk. You press ↓ and ↑ to highlight the desired graph, and then press **Spacebar** to select it. (The same command will *unselect* a previously selected graph.)

To select and preview a graph

Press	**S** (for *Select*) to display a list of the graphs on the disk
Press	**↓** (if necessary) to highlight *PV2,* and then press **Spacebar** to place a pound sign (*#*) in front of it
Press	**Graph (F10** on the IBM PC) to preview the graph (if your computer is equipped to display graphics)
Press	**Spacebar** to clear the graph, and then press **Enter** to return to the menu

Step 4 Select a Font and Size. You have to select a font for the labels before you can print a graph. We will also select a small size for the graph so that it prints more quickly.

To select a font

Press	**OF** (for *Options*, *Font*), and the prompt reads *1 2*
Press	**1** and then highlight any font
Press	**Spacebar** to select the highlighted font, and then press **Enter**
Press	**F** (for *Font*), and the prompt reads *1 2*
Press	**2** and then highlight any font
Press	**Spacebar** to select the highlighted font, and then press **Enter**

To specify the size of the graph

Press	**SM** (for *Size*, *Manual*)
Press	**W** (for Width), and the prompt reads *Enter number of inches of width*
Type	**7** and then press **Enter**
Press	**H** (for Height), and the prompt reads *Enter number of inches of height*
Type	**5** and then press **Enter**
Press	**Q** (for *Quit*) three times to return to the Graph menu

Step 5 Print the Graph. Now that you have selected a graph, specified fonts, and specified the graph's size, you can print it. Be sure your printer is on and has paper in it.

To print the graph

Press	**AG** (for *Align*, *Go*)

Result. After a few moments, the graph begins printing. When it is finished printing, press **QY** (for *Quit*, *Yes*) to return to the Lotus Access System. Insert the Lotus 1-2-3 System Disk in drive A, and then press **1** (for *1-2-3*) to return to the worksheet. When the copyright screen appears, press any key to load AUTO123.

EXERCISE 4
MASTER BUDGETING

In this exercise, you solve cases that are based on the principles of master budgeting discussed in Chapter 6 of *Cost Accounting: A Managerial Emphasis* (8th edition). This exercise assumes that the template MSTBUDGT is displayed on the screen. (If you need help on retrieving files, refer to Exercise 1 or Appendix C.)

UNDERSTANDING THE TEMPLATE

The MSTBUDGT template has three parts (Figure 24 and Table 5). It is based on the budget schedules shown in Exhibit 6-1 in *Cost Accounting: A Managerial Emphasis* (8th edition). Data from Halifax Engineering, the example used in your text, has already been entered in the template as a starting point for analyzing the cases.

Part 1. Variables is the only part of the template in which you enter data for the cases. The starting data for Case 1 has been entered. You enter data for the other cases at the appropriate time.

Part 2. Budgeting contains all the schedules that make up the master budget. This part is not broken into "screens" between which you can quickly page like the other templates you have used. Press ↓ to scroll down through this part to see the sales, production, direct materials, direct material purchases, direct labor, factory overhead, ending-inventory, cost of goods sold, and selling and administrative budgets. All the numbers in these schedules have been entered for you to save you time.

> **TIP ON GETTING AROUND PART 2 OF MSTBUDGT**
>
> To help you get around this part, each schedule has been named with an abbreviation, 1, 2, 3A, 3B, and so on. To display a specific schedule, press **Goto** (**F5** on the IBM PC), type the abbreviation, and then press **Enter**.

Part 3. Exhibits contains two financial statements. All the figures in both statements (in the protected cells) are calculated based on data entered in Parts 1 and 2. Changes in the values in any of the unprotected cells in Parts 1 and 2 result in an automatic recalculation of these two statements.

FIGURE 24 The MSTBUDGT Template.

```
              A                B              C                D          E            F
 1  PART 1. VARIABLES
 2  ==================== ======== ==================== ========= =========== ============
 3  Case 1. Altering quantity sold
 4
 5    Quantity Product R sold....................................:                  5,000
 6    Quantity Product HD sold...................................:                  1,000
 7
 8  Case 2. Altering inventory policy
 9
10    Days raw material inventory - Material 111.................: #________
11    Days raw material inventory - Material 112.................: #________
12    Days finished goods inventory - Product R..................: #________
13    Days finished goods inventory - Product HD.................: #________
14
15  Case 3. Altering fixed costs
16
17    Additional fixed selling costs.............................: $________
18    Additional fixed production costs..........................: $________
19
20  ==================== ======== ==================== ========= =========== ============
21  PART 2. BUDGETING
22  ==================== ======== ==================== ========= =========== ============
23  SCHEDULE 1. SALES BUDGET                                  Units     Selling       Total
24                                                                      price       Sales
25                                                        --------- ----------- ------------
26  Product R - regular aircraft parts                      5,000        $600   3,000,000
27  Product HD - heavy duty aircraft parts                  1,000        $800     800,000
28                                                                              -----------
29                                                                             $3,800,000
30  -------------------- -------- -------------------- --------- ----------- ------------
31  SCHEDULE 2. PRODUCTION BUDGET                                   [----Products----]
32                                                                          R          HD
33                                                                  ----------- ------------
34  Budgeted sales (Schedule 1)                                           5,000       1,000
35  Target ending finished-goods inventory                                1,100          50
36                                                                  ----------- ------------
37    Total needs                                                         6,100       1,050
38  Less beginning finished-goods inventory                                 100          50
39                                                                  ----------- ------------
40    Units to be produced                                                6,000       1,000
41  -------------------- -------- -------------------- --------- ----------- ------------
42  SCHEDULE 3A. DIRECT MATERIALS IN KILOGRAMS AND DOLLARS
43
44                                                         [----Materials---]
45                                                              111         112        Total
46                                                        --------- ----------- ------------
47  Product R unit requirement                                    12           6
48  Product HD unit requirement                                   12           8
49  Product R total requirement                               72,000      36,000
50  Product HD total requirement                              12,000       8,000
51                                                        --------- -----------
52  Total direct material usage                               84,000      44,000
53  Price per kilogram                                         $7.00      $10.00
54                                                        --------- -----------
55  Total direct cost of materials used                     $588,000    $440,000  $1,028,000
56  -------------------- -------- -------------------- --------- ----------- ------------
```

FIGURE 24 (Cont.)

SCHEDULE 3B. DIRECT MATERIAL-PURCHASES BUDGET

	[----Materials---]		
	111	112	Total
Target end direct-material inventory in k	8,000	2,000	
Target kilograms used in production	84,000	44,000	
Total needs	92,000	46,000	
Less begin. direct-material inventory in k	7,000	6,000	
Kilograms to be purchased	85,000	40,000	
Price per kilogram	$7.00	$10.00	
Purchase cost	$595,000	$400,000	$995,000

SCHEDULE 4. DIRECT LABOR BUDGET

	Units produced	Direct-labor hrs/unit	Total hours	Hourly rate	Total budget
Product R	6,000	4	24,000	$20.00	$480,000
Product HD	1,000	6	6,000	$20.00	120,000
Total			30,000		$600,000

SCHEDULE 5. MANUFACTURING OVERHEAD BUDGET

Supplies	$90,000	
Indirect labor	200,000	
Payroll fringe costs	320,000	
Power-variable portion	90,000	
Maintenance-variable portion	70,000	
Total variable overhead		
		$770,000
Depreciation	$230,000	
Property taxes	50,000	
Property insurance	10,000	
Supervision	100,000	
Power-fixed portion	20,000	
Maintenance-fixed portion	20,000	
Total fixed overhead		430,000
Total manufacturing overhead		$1,200,000
Per direct-labor hour		$40.00

SCHEDULE 6. ENDING INVENTORY BUDGET

Direct materials:	Kilograms	$/K	Total
Material 111	8,000	$7.00	$56,000
Material 112	2,000	$10.00	20,000
Total			$76,000

FIGURE 24 (Cont.)

	A	B	C	D	E	F
114	Finished goods:			Units	$/unit	Total
115				---------	-----------	-----------
116	Product R			1,100	$384.00	$422,400
117	Product HD			50	$524.00	26,200
118						-----------
119	Total					$448,600
120						-----------
121	Total ending inventory					$524,600
122						===========
123		$/K or	[----Product R---]		[---Product HD---]	
124		hour/out	Inputs	Amount	Inputs	Amount
125		--------	------------------	---------	-----------	-----------
126	Material 111	$7.00	12	$84.00	12	$84.00
127	Material 112	$10.00	6	60.00	8	80.00
128	Direct labor	$20.00	4	80.00	6	120.00
129	Manufacturing overhead	$40.00	4	160.00	6	240.00
130				---------		-----------
131	Total			$384.00		$524.00
132	--------------------	--------	------------------	---------	-----------	-----------
133	SCHEDULE 7. COST OF GOODS SOLD BUDGET					
134						
135	Finished goods inventory 12/31/19_1					$64,600
136	Direct materials used				1,028,000	
137	Direct labor				600,000	
138	Manufacturing overhead				1,200,000	
139	Cost of goods manufactured					$2,828,000
140						-----------
141	Cost of goods available for sale					$2,892,600
142	Finished goods inventory 12/31/19_2					448,600
143						-----------
144	Cost of goods sold					$2,444,000
145	--------------------	--------	------------------	---------	-----------	-----------
146	SCHEDULE 8. PRODUCT DEVELOPMENT SCHEDULE					
147						
148	Product development salaries					$105,000
149	Prototype and material cost					31,000
150						-----------
151	Total product development costs					$136,000
152	--------------------	--------	------------------	---------	-----------	-----------
153	SCHEDULE 9. MARKETING BUDGET					
154						
155	Advertising and Promotion					$30,000
156	Marketing salaries					130,000
157	Travel					40,000
158						-----------
159	Total marketing costs					$200,000
160	--------------------	--------	------------------	---------	-----------	-----------
161	SCHEDULE 10. DISTRIBUTION BUDGET					
162						
163	Distribution salaries and wages					$60,000
164	Rent, power, maintenance, and taxes					40,000
165						-----------
166	Total distribution costs					$100,000
167	--------------------	--------	------------------	---------	-----------	-----------

FIGURE 24 (Cont.)

	A	B	C	D	E	F
168	SCHEDULE 11. CUSTOMER SERVICE BUDGET					
169						
170	Customer service salaries					$40,000
171	Travel and miscellaneous					20,000
172						------------
173	Total customer service costs					$60,000
174	--------------------	--------	------------------	---------	-----------	------------
175	SCHEDULE 12. ADMINISTRATIVE BUDGET					
176						
177	Clerical wages					$80,000
178	Office costs and supplies					75,000
179	Office salaries					165,000
180	Miscellaneous					54,000
181						------------
182	Total administrative costs					$374,000
183	--------------------	--------	------------------	---------	-----------	------------
184						
185	PART 3. EXHIBITS					
186	====================	========	==================	=========	===========	============
187	A. BUDGETED INCOME STATEMENT					
188						
189	Sales					$3,800,000
190	Cost of goods sold					2,444,000
191						------------
192	Gross margin					1,356,000
193	Product development					136,000
194	Marketing, distribution, and customer service					360,000
195	Administrative					374,000
196						------------
197	Operating income					$486,000
198						============
199	B. BUDGETED BALANCE SHEET					
200						
201	ASSETS					
202	Current assets					
203	Cash				$35,000	
204	Accounts receivable				316,667	
205	Direct materials				76,000	
206	Finished goods				448,600	$876,267
207					------------	
208	Fixed assets					
209	Land				200,000	
210	Buildings and equipment				2,600,000	
211	Accumulated depreciation				-920,000	1,880,000
212					------------	------------
213	Total assets					$2,756,267
214						============
215	LIABILITIES					
216	Accounts payable				$82,917	
217	Income taxes payable				194,400	
218	Notes payable				73,750	$351,067
219	STOCKHOLDER'S EQUITY				------------	
220	Common stock				350,000	
221	Retained income				2,055,200	2,405,200
222					------------	------------
223	Total liabilities & equities					$2,756,267
224						============

Table 5
THE MSTBUDGT TEMPLATE

Rows	Description
Part 1. Variables	
5-6	You enter or change the quantities sold on these rows when solving Case 1.
10-13	You change target ending inventory figures here when solving Case 2.
17-18	You change fixed costs here when solving Case 3.
Part 2. Budgeting	
23-29	Schedule 1, Sales Budget.
31-40	Schedule 2, Production Budget.
42-55	Schedule 3A, Direct Materials (in kilograms and dollars).
57-71	Schedule 3B, Direct Material-Purchases Budget.
73-81	Schedule 4, Direct Labor Budget.
83-103	Schedule 5, Manufacturing Overhead Budget.
105-131	Schedule 6, Ending Inventory Budget.
133-144	Schedule 7, Cost of Goods Sold Budget.
146-151	Schedule 8, Product Development Budget.
153-159	Schedule 9, Marketing Budget.
161-166	Schedule 10, Distribution Budget.
168-173	Schedule 11, Customer Service Budget.
175-182	Schedule 12, Administrative Budget.
Part 3. Exhibits	
187-197	A budgeted income statement.
199-223	A budgeted balance sheet.

CASE 1 HALIFAX ENGINEERING — PART I

This case demonstrates how the schedules of the master budget interrelate. The master budget is described in Chapter 6 of your text and illustrated in Exhibit 6-1.

The template automates the calculations for you so that you can concentrate on the changed input variables and the resulting changes in the master budget schedules. Data from Halifax Engineering, which is the example used in the text, have already been entered in the template. All you must do is follow the case requirements to change selected input variables and observe and record the changes in the master budget schedules.

This case specifically examines the impact of sales quantities on the master budget schedules.

REQUIREMENTS

A. Increase the sales of Product R from 5,000 units to 5,500 units. What will the following items be: (1) production of Product R, (2) direct materials purchased — Material 111 and Material 112, (3) direct labor hours, (4) total factory overhead, (5) total selling and administrative expenses, and (6) operating income? Explain why these items changed.

B. Return Product R sales to 5,000 units, and increase Product HD sales to 1,100 units. What will the following items be: (1) production of Product HD, (2) direct materials purchased — Material 111 and Material 112, (3) direct labor hours, (4) total factory overhead, (5) total selling and administrative expenses, and (6) operating income? Explain why these items changed.

C. Return Product HD sales to 1,000 units. Shift 200 units of sales from Product R to Product HD (sales of 4,800 units for Product R and 1,200 units for Product HD). What will the following items be: (1) production of Product R and Product HD, (2) direct materials purchased — Material 111 and Material 112, (3) direct labor hours, (4) total factory overhead, (5) total selling and administrative expenses, and (6) operating income? Explain why these items changed.

D. Assume sales of Product R are 4,900 units and Product HD are 950 units. What will the following items be: (1) production of Product R and Product HD, (2) direct materials purchased — Material 111 and Material 112, (3) direct labor hours, (4) total factory overhead, (5) total selling and administrative expenses, and (6) operating income? Explain why these items changed.

Name:________________________

CASE 1. HALIFAX ENGINEERING — PART I SOLUTION
Template: MSTBUDGT

DATA ANALYSIS SUMMARY (attach model printouts as supporting documents):

Halifax Engineering
Master Budget — Volume Changes

REQUIREMENT	PRODUCTION		MATERIALS PURCHASED	
	R	HD	111	112
Text example	6,000	1,000	$595,000	$400,000
A		1,000		
B	6,000			
C				
D				

REQUIREMENT	LABOR HOURS	TOTAL OVERHEAD	TOTAL SELLING	OPERATING INCOME
Text example	30,000	$1,200,000	$870,000	$486,000
A				
B				
C				
D				

DISCUSSION

CASE 2 HALIFAX ENGINEERING — PART II

This case demonstrates how the schedules of the master budget interrelate. The master budget is described in Chapter 6 of your text and illustrated in Exhibit 6-1.

The template automates the calculations for you so that you can concentrate on the changed input variables and the resulting changes in the master budget schedules. Data from the Halifax Engineering, which is the example used in the text, have already been entered in the template. All you must do is follow the case requirements to change selected input variables and observe and record the changes in the master budget schedules.

This case specifically examines the impact of inventory policies on the master budget schedules. We assume we need raw materials inventory so that we can purchase and ship in economical quantities and have enough raw material so that production is not interrupted if a shipment is delayed. We also assume we need finished goods inventory for unexpected customer orders and to keep production uniform if sales are seasonal. Recently, these assumptions about inventory policies have been severely questioned. There are reports of Japanese and U.S. companies adopting just-in-time (JIT) delivery of raw materials, which reduces the level of inventory without production interruptions. There are also reports of companies with computer links to their major customers so that they can schedule production to meet their customers' needs. This reduces the requirements for finished goods inventory. A convenient way of expressing an inventory policy is in terms of days inventory on hand (or inventory days). Thus, if we have 30 days of raw material this means that based on average usage, it would take 30 days of production to use the raw material we currently have on hand.

Although not stated, the text example contains the following inventory days in the related targeted ending inventories:

Material 111	34.3 days	Product R	79.2 days
Material 112	16.4 days	Product HD	18.0 days

These are the days inventory that are used when the input cells for days inventory in Part 1 of the template contain all blanks. Once you enter any number in one of these input cells, that number is used to calculate the related targeted ending inventory.

REQUIREMENTS

A. Decrease the raw materials inventory to 9 days for Material 111 and 4 days for Material 112. What will the following items be: (1) units in ending inventory of direct materials — Material 111 and Material 112, (2) ending raw material inventory on the balance sheet, and (3) accounts and notes payable on the balance sheet? Explain why these items changed.

B. Use the template menu's Clear command to clear the template's input cells. Decrease finished goods inventory to 40 days for Product R and 10 days for Product HD. What will the following items be: (1) units in ending inventory of Product R and Product HD, (2) ending finished goods inventory on the balance sheet, and (3) accounts and notes payable on the balance sheet? Explain why these items changed.

C. What other potential savings or advantages might result from lower inventory levels?

Name:________________________

CASE 2. HALIFAX ENGINEERING — PART II SOLUTION
Template: MSTBUDGT

DATA ANALYSIS SUMMARY (attach model printouts as supporting documents):

Halifax Engineering
Master Budget — Inventory Changes

REQUIREMENT	E.I.- RAW MATERIAL			E.I.- FIN. GOODS			PAYABLES	
	UNITS		TOTAL	UNITS		TOTAL		
	111	112	$	R	HD	$	ACCOUNTS	NOTES
Text example	8,000	2,000	$76,000	1,100	50	$448,600	$82,917	$73,750
A				NR	NR	NR		
B	NR	NR	NR					

NR = not required.

DISCUSSION

CASE 3 HALIFAX ENGINEERING — PART III

This case demonstrates how the schedules of the master budget interrelate. The master budget is described in Chapter 6 of your text and illustrated in Exhibit 6-1.

The template automates the calculations for you so that you can concentrate on the changed input variables and the resulting changes in the master budget schedules. Data from the Halifax Engineering, which is the example used in the text, have already been entered in the template. All you must do is follow the case requirements to change selected input variables and observe and record the changes in the master budget schedules.

This case specifically examines the impact on the master budget schedules of additional fixed costs caused by volume increases. As you will recall from Chapter 2, fixed costs remain fixed only within the relevant range of activity. If an increase in volume moves production beyond the relevant range, the company may have to budget additional fixed costs.

REQUIREMENTS

A. Increase the sales of Product R to 5,250 units and Product HD to 1,050 units in Part 1 of the template. What will the following items be: (1) total factory overhead, (2) total selling and administrative expenses, and (3) operating income? Explain why these items changed.

B. Suppose the increase in units sold described in Requirement A required an additional $25,000 in fixed promotional costs. (Enter the $25,000 in Part 1 of the template.) What will the following items be: (1) total factory overhead, (2) total selling and administrative expenses, and (3) operating income? Is the volume increase still desirable?

C. Suppose the increase in units sold described in Requirement A will also require the company to produce on a second shift with a resulting increase in fixed factory overhead of $40,000. This fixed cost increase is in addition to the $25,000 in promotional costs in Requirement B. What will the following items be: (1) total factory overhead, (2) total selling and administrative expenses, and (3) operating income? Is the volume increase still desirable?

D. Examine the balance sheet impact of this volume increase and the related fixed cost increases mentioned in Requirements B and C. What will the following items be: (1) accounts receivable, (2) ending inventory raw materials, (3) ending inventory finished goods, and (4) notes payable. Explain why these items changed.

Name:______________________

CASE 3. HALIFAX ENGINEERING — PART III SOLUTION
Template: MSTBUDGT

DATA ANALYSIS SUMMARY (attach model printouts as supporting documents):

Halifax Engineering
Master Budget — Volume & Fixed Cost Increases

REQUIREMENT	UNITS SOLD		TOTAL	TOTAL	OPERATING
	R	HD	OVERHEAD	SELLING	INCOME
Text example	5,000	1,000	$1,200,000	$870,000	$486,000
A	5,250	1,050			
B	5,250	1,050			
C	5,250	1,050			

Halifax Engineering
Balance Sheet Changes

REQUIREMENT	ACCOUNTS	ENDING INVENTORY		NOTES
	RECEIVABLE	RAW MAT.	FIN. GDS.	PAYABLE
Text example	$316,667	$76,000	$448,600	$73,750
D				

DISCUSSION

TUTORIAL 4 USING A DATA TABLE

You can explore what-ifs by changing values one at a time and watching the results. Data tables, however, offer a faster way to do this. You can use the Data Table 1 commands to run a series of values through a cell on the worksheet and capture the results from a number of other cells. Let's see how you would do this to change the quantity of Product HD in cell F6 from 0 to 1100 in increments of 100 and capture the resulting output from three cells on the budgeted income statement.

Step 1 Retrieve the MSTBUDGT Template. Retrieve the template if it isn't already on the screen. Use the Clear command on the template menu to return the template to its original state so that your results match ours. Press **/WGPD** (for *Worksheet*, *Global*, *Protection*, *Disable*) to turn off protection.

Step 2 Enter the Series of Values to Be Run Through Cell F6. You want to run a series of values from 0 to 1100 units through cell F6 to see how key figures on the budgeted income statement are affected by changes in the quantity of Product HD that are sold.

To enter a series of numbers

Goto	G1, type **`Start`** and then press **Enter**
Goto	H1, type `0` and then press **Enter**
Goto	G2, type `Increment` and then press **Enter**
Goto	H2, type `100` and then press **Enter**
Goto	G6, type `+H1` and then press **Enter**
Goto	G7, type `+G6+H$2` and then press **Enter**
Press	/C (for *Copy*), and a prompt asks you to enter the range to copy FROM
Press	Enter, and a prompt asks you to enter the range to copy TO
Type	`G8.G17` and then press **Enter**

Result. Column G fills with numbers from 0 to 1100 in steps of 100.

Step 3 Enter the Cells from Which You Want to Capture the Output. Your program allows you to capture the output from up to eight cells. Here you capture the output from sales (cell F166), gross margin (cell F169), and operating income (cell F172).

To enter the output cells

Goto	H5, type **`+F166`** and then press **Enter**
Goto	I5, type `+F169` and then press **Enter**
Goto	J5, type `+F172` and then press **Enter**

To format the output cells as text

Press	**/RFT** (for *Range*, *Format*, *Text*), and the prompt reads *Enter range to format*
Type	**`H5.J5`** and then press **Enter**

Result. The text format displays the cell references as text so that you can see what cells are referred to.

Step 4 Run the Data Table. Now let's run the data table to see what happens.

To run the data table

Press **/DT1** (for *Data, Table, 1*), and the prompt asks you to enter the table range

Type **G5.J17** and then press **Enter**. The prompt asks you to enter the input cell 1

Type **F6** and then press **Enter**

FIGURE 25 The Data Table 1. After you create and run the data table, a range of values is automatically generated.

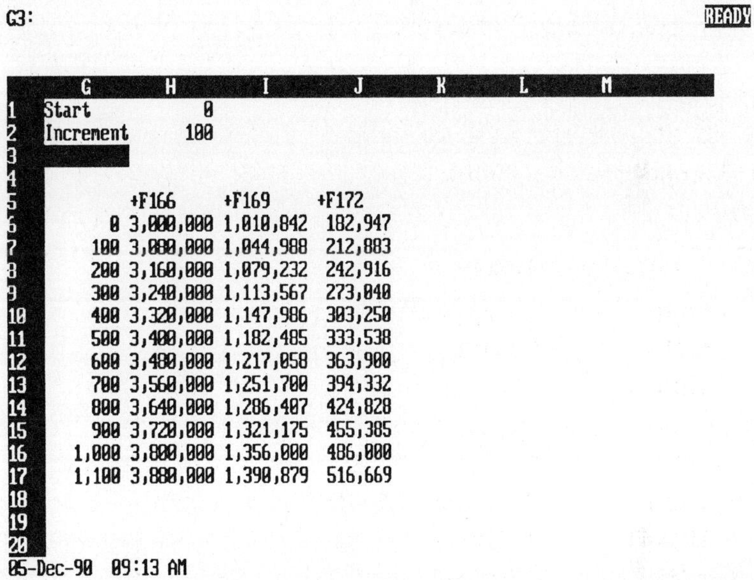

G3: READY

	G	H	I	J	K	L	M
1	Start	0					
2	Increment	100					
3							
4							
5		+F166	+F169	+F172			
6	0	3,000,000	1,010,842	182,947			
7	100	3,080,000	1,044,988	212,883			
8	200	3,160,000	1,079,232	242,916			
9	300	3,240,000	1,113,567	273,040			
10	400	3,320,000	1,147,986	303,250			
11	500	3,400,000	1,182,485	333,538			
12	600	3,480,000	1,217,058	363,900			
13	700	3,560,000	1,251,700	394,332			
14	800	3,640,000	1,286,407	424,828			
15	900	3,720,000	1,321,175	455,385			
16	1,000	3,800,000	1,356,000	486,000			
17	1,100	3,880,000	1,390,879	516,669			
18							
19							
20							

05-Dec-90 09:13 AM

Result. The mode indicator reads *WAIT* while the program enters each of the values in column A into cell F6, recalculates the worksheet, and stores the results from the output cells in the data table. In a moment, the calculated results are displayed (Figure 25). If you compare the results for 1000 units on the table, you will see that sales (H16) are $3,800,000, gross margin (I16) is $1,356,000, and operating income (J16) is $486,000. These results match those on the respective cells on the income statement. The other rows on the data table show how these outputs are affected by changes in the quantity of Product HD that is sold.

Step 5 Run the Data Table Again. Data tables remember their settings. To use them again, just change variables on the template, and then press **Table** (**F8** on the IBM PC). Here, change the increment to change the range of values in column A, and then run the table to see the new results.

To change the input values

Goto H2, type **50** and then press **Enter**

To run the table again

Press **F8**

Result. The mode indicator reads *WAIT* while the program enters each of the values in column A into cell F6, recalculates the worksheet, and stores the results from the output cells in the data table. In a moment, the calculated results are displayed. These results show the effect of the lower quantities of Product HD on the income statement.

To save your changes, press **Alt-M** to display the template menu, and then press **S** (for *Save*).

EXERCISE 5
CASH BUDGETING

In this exercise, you solve cases that are based on the principles of cash budgeting discussed in the appendix of Chapter 6 of *Cost Accounting: A Managerial Emphasis* (8th edition). This exercise assumes that the template CASHFLOW is displayed on the screen. (If you need help on retrieving files, refer to Exercise 1 or Appendix C.)

TIP ON ENTERING PRINTER SETUP CODES

If you are not using an Epson or Epson-compatible printer, and you did not enter setup codes for your printer in Exercise 1 or in another exercise, you have to enter them before you can print this template. You can do so when the CASHFLOW template is displayed on the screen. Begin by pressing **Alt-M** to display the template menu. Then press **P** (for *Print*) and **S** (for *Setup*). The prompt at the top of the screen reads *Enter setup code for compressed type:*. Type `\015` (or whatever code is correct for your printer), and then press **Enter**. The prompt reads *Enter setup code for normal type:*. Type `\018` (or whatever code is correct for your printer), and then press **Enter**. The disk drive spins, and a message asks you to wait while a file containing the codes is saved onto the disk. In a moment, the AUTO123 template menu reappears. You will not have to enter these codes again unless you change printers.

EXPLORING THE TEMPLATE

The CASHFLOW template has six parts (Figure 26 and Table 6). You use the Data/Cases command on the template menu to enter data into Parts 1 through 3. Then, press **Calc** (**F9** on the IBM PC), and the rest of the template calculates the cash budget automatically. Since recalculation is set to manual, you will have to press **Calc** whenever you change a case assumption to calculate the impact on the cash budget.

Parts 4 through 6 are wider than the screen, so you have to scroll sideways to see all of them. Press **Tab** to scroll to the right one screen. Hold down **Shift** and press **Tab** to move back to the left one screen. To keep the row labels in column A from scrolling off the screen when you press **Tab**, see the tutorial at the end of this exercise or refer to Appendix I, "Using Fixed Titles."

Part 1. Sales & Cash Receipts Assumptions is where you enter assumptions about sales and cash receipts.

Part 2. Purchases & Cash Payments Assumptions is where you enter assumptions about purchases and cash payments.

Part 3. Short-Term Financing Assumptions is where you enter assumptions about your cash (or money market) balance, notes payable, and opening accounts receivable and accounts payable.

Part 4. Sales & Cash Receipts Budget calculates the sales and cash receipts portion of the cash budget based on the data you supply in Parts 1 through 3.

Part 5. Purchases & Cash Payments Budget calculates the purchases and cash payments portion of the cash budget based on the data you supply in Parts 1 through 3.

Part 5. Purchases & Cash Payments Budget calculates the purchases and cash payments portion of the cash budget based on the data you supply in Parts 1 through 3.

Part 6. Cash Budget Summary summarizes the cash receipts from Part 4 and the cash payments from Part 5 and then calculates the budgeted short-term financing activity.

FIGURE 26 The CASHFLOW Template.

```
                        A                B   C        D         E          F
 1 PART 1 SALES & CASH RECEIPTS ASSUMPTIONS
 2 ============================ = ======== ======== ======== ==========
 3 Enter total annual sales.................................: $________
 4 Enter sales pattern                    Jan-X2 _______%   Jul-X2  _______%
 5                                        Feb-X2 _______%   Aug-X2  _______%
 6                                        Mar-X2 _______%   Sep-X2  _______%
 7                                        Apr-X2 _______%   Oct-X2  _______%
 8                                        May-X2 _______%   Nov-X2  _______%
 9                                        Jun-X2 _______%   Dec-X2  _______%
10                                     Total sales                        NA
11 ---------------------------- - -------- -------- -------- ----------
12 % cash sales                                                   _______%
13 % collected in discount period (10 days)  Discount: ______%    _______%
14 % collected within 30 days                                     _______%
15 % collected within 60 days                                     _______%
16 % collected within 90 days                                     _______%
17                                     Total cash receipts                NA
18 ---------------------------- - -------- -------- -------- ----------
19 Enter first qtr/next period increase over same qtr this year   _______%
20 ============================ = ======== ======== ======== ==========
21 PART 2. PURCHASES & CASH PAYMENTS ASSUMPTIONS
22 ============================ = ======== ======== ======== ==========
23 Enter cost of goods sold (%)                                   _______%
24 Enter inventory purchased lead time in months (0 to 3)         _______#
25 ---------------------------- - -------- -------- -------- ----------
26 Enter cash purchases                                           _______%
27 % paid within discount period (10 days)   Discount: ______%    _______%
28 % paid within 30 days                                          _______%
29 % paid within 60 days                                          _______%
30                                     Total purchases                    NA
31 ---------------------------- - -------- -------- -------- ----------
32 Enter variable operating expense (%)                           _______%
33 Enter fixed operating expense per month                        $_______
34 Enter quarterly tax payments                                   $_______
35 ---------------------------- - -------- -------- -------- ----------
36 Enter advertising                    Quarter   Amount   Quarter    Amount
37 expense pattern                      ------- -------- -------- ----------
38                                        First $______    Third  $_______
39                                       Second $______   Fourth  $_______
40 ============================ = ======== ======== ======== ==========
41 PART 3. SHORT TERM FINANCING ASSUMPTIONS
42 ============================ = ======== ======== ======== ==========
43
44 Cash balance/money market:
45
46 Enter minimum cash balance desired                             $_______
47 Enter opening cash/money market balance                        $_______
48 Enter money market interest rate                                ______%
49
50 Notes payable:
51
52 Enter opening notes payable balance                            $_______
53 Enter notes payable interest rate                               ______%
54
55 Opening A/R and A/P:
56
57 Enter opening accounts receivable                              $_______
58 Enter opening accounts payable                                 $_______
59
60 ============================ = ======== ======== ======== ==========
```

FIGURE 26 (Cont.)

	A	B	C	D	E	F	G	H	I
61	PART 4. SALES & CASH RECEIPTS BUDGET								
62	============================	=	==========	==========	==========	==========	==========	==========	==========
63	ITEM: Date:	|	Jan-X2	Feb-X2	Mar-X2	Apr-X2	May-X2	Jun-X2	Jul-X2
64	----------------------------	-	----------	----------	----------	----------	----------	----------	----------
65	Net sales	|	$0	$0	$0	$0	$0	$0	$0
66	============================	=	==========	==========	==========	==========	==========	==========	==========
67	[CASH RECEIPTS]	|							
68	Cash sales	|	$0	$0	$0	$0	$0	$0	$0
69	% collected in discount period	|	$0	$0	$0	$0	$0	$0	$0
70	% collected within 30 days	|	$0	$0	$0	$0	$0	$0	$0
71	% collected within 60 days	|	$0	$0	$0	$0	$0	$0	$0
72	% collected within 90 days	|	$0	$0	$0	$0	$0	$0	$0
73	[.........................]	|	$______	$______	$______	$______	$______	$______	$______
74	[.........................]	|	$______	$______	$______	$______	$______	$______	$______
75	[.........................]	|	$______	$______	$______	$______	$______	$______	$______
76	[.........................]	|	$______	$______	$______	$______	$______	$______	$______
77	[.........................]	|	$______	$______	$______	$______	$______	$______	$______
78	----------------------------	-	----------	----------	----------	----------	----------	----------	----------
79	Total cash receipts	|	$0	$0	$0	$0	$0	$0	$0
80	============================	=	==========	==========	==========	==========	==========	==========	==========
81	PART 5. PURCHASES & CASH PAYMENTS BUDGET								
82	============================	=	==========	==========	==========	==========	==========	==========	==========
83	ITEM: Date:	|	Jan-X2	Feb-X2	Mar-X2	Apr-X2	May-X2	Jun-X2	Jul-X2
84	----------------------------	-	----------	----------	----------	----------	----------	----------	----------
85	Total purchases	|	$0	$0	$0	$0	$0	$0	$0
86	============================	=	==========	==========	==========	==========	==========	==========	==========
87	[CASH PAYMENTS]								
88	Cash purchases	|	$0	$0	$0	$0	$0	$0	$0
89	% paid within discount period	|	$0	$0	$0	$0	$0	$0	$0
90	% paid within 30 days	|	$0	$0	$0	$0	$0	$0	$0
91	% paid within 60 days	|	$0	$0	$0	$0	$0	$0	$0
92	----------------------------	-	----------	----------	----------	----------	----------	----------	----------
93	Total inventory payments	|	$0	$0	$0	$0	$0	$0	$0
94	Variable operating expenses	|	$0	$0	$0	$0	$0	$0	$0
95	Fixed operating expenses	|	$______	$______	$______	$______	$______	$______	$______
96	Quarterly tax payments	|	$0	----	----	$0	----	----	$0
97	Advertising expenses	|	$0	$0	$0	$0	$0	$0	$0
98	----------------------------	-	----------	----------	----------	----------	----------	----------	----------
99	Total cash payments	|	$0	$0	$0	$0	$0	$0	$0
100	============================	=	==========	==========	==========	==========	==========	==========	==========
101	PART 6. CASH BUDGET SUMMARY								
102	============================	=	==========	==========	==========	==========	==========	==========	==========
103	ITEM: Date:	|	Jan-X2	Feb-X2	Mar-X2	Apr-X2	May-X2	Jun-X2	Jul-X2
104	----------------------------	-	----------	----------	----------	----------	----------	----------	----------
105	Net monthly cash flow	|	$0	$0	$0	$0	$0	$0	$0
106	Opening cash/money market bal.	|	NA	NA	NA	NA	NA	NA	NA
107	Interest earned	|	NA	NA	NA	NA	NA	NA	NA
108	Interest paid	|	$0	$0	$0	$0	$0	$0	$0
109	----------------------------	-	----------	----------	----------	----------	----------	----------	----------
110	Cash available	|	$0	$0	$0	$0	$0	$0	$0
111	----------------------------	-	----------	----------	----------	----------	----------	----------	----------
112	Borrowing activity	|							
113	Opening notes payable	|	$0	$0	$0	$0	$0	$0	$0
114	Borrowing	|	NA	NA	NA	NA	NA	NA	NA
115	Notes repaid	|	NA	NA	NA	NA	NA	NA	NA
116	----------------------------	-	----------	----------	----------	----------	----------	----------	----------
117	Closing notes payable balance	|	NA	NA	NA	NA	NA	NA	NA
118	============================	=	==========	==========	==========	==========	==========	==========	==========
119	End. cash/marketable securitie	|	NA	NA	NA	NA	NA	NA	NA
120	============================	=	==========	==========	==========	==========	==========	==========	==========
121	Graph X-axis	|	1	2	3	4	5	6	7
122	============================	=	==========	==========	==========	==========	==========	==========	==========

FIGURE 26 (Cont.)

	J	K	L	M	N	O	P	Q
61						- 1st QTR NEXT YEAR		
62	==========	==========	==========	==========	===========	=========	=========	=========
63	Aug-X2	Sep-X2	Oct-X2	Nov-X2	Dec-X2	Jan-X3	Feb-X3	Mar-X3
64	----------	----------	----------	----------	-----------	---------	---------	---------
65	$0	$0	$0	$0	$0	$0	$0	$0
66	==========	==========	==========	==========	===========	=========	=========	=========
67								
68	$0	$0	$0	$0	$0			
69	$0	$0	$0	$0	$0			
70	$0	$0	$0	$0	$0			
71	$0	$0	$0	$0	$0			
72	$0	$0	$0	$0	$0			
73	$________	$________	$________	$________	$________			
74	$________	$________	$________	$________	$________			
75	$________	$________	$________	$________	$________			
76	$________	$________	$________	$________	$________			
77	$________	$________	$________	$________	$________			
78	----------	----------	----------	----------	-----------			
79	$0	$0	$0	$0	$0			
80	==========	==========	==========	==========	===========	=========	=========	=========
81								
82	==========	==========	==========	==========	===========			
83	Aug-X2	Sep-X2	Oct-X2	Nov-X2	Dec-X2			
84	----------	----------	----------	----------	-----------			
85	$0	$0	$0	$0	$0			
86	==========	==========	==========	==========	===========			
87								
88	$0	$0	$0	$0	$0			
89	$0	$0	$0	$0	$0			
90	$0	$0	$0	$0	$0			
91	$0	$0	$0	$0	$0			
92	----------	----------	----------	----------	-----------			
93	$0	$0	$0	$0	$0			
94	$0	$0	$0	$0	$0			
95	$________	$________	$________	$________	$________			
96	----	----	$0	----	----			
97	$0	$0	$0	$0	$0			
98	----------	----------	----------	----------	-----------			
99	$0	$0	$0	$0	$0			
100	==========	==========	==========	==========	===========			
101								
102	==========	==========	==========	==========	===========			
103	Aug-X2	Sep-X2	Oct-X2	Nov-X2	Dec-X2			
104	----------	----------	----------	----------	-----------			
105	$0	$0	$0	$0	$0			
106	NA	NA	NA	NA	NA			
107	NA	NA	NA	NA	NA			
108	$0	$0	$0	$0	$0			
109	----------	----------	----------	----------	-----------			
110	$0	$0	$0	$0	$0			
111	----------	----------	----------	----------	-----------			
112								
113	$0	$0	$0	$0	$0			
114	NA	NA	NA	NA	NA			
115	NA	NA	NA	NA	NA			
116	----------	----------	----------	----------	-----------			
117	NA	NA	NA	NA	NA			
118	==========	==========	==========	==========	===========			
119	NA	NA	NA	NA	NA			
120	==========	==========	==========	==========	===========			
121	8	9	10	11	12			
122	==========	==========	==========	==========	===========			

Table 6
THE CASHFLOW TEMPLATE

Rows	Description
Part 1. Sales & Cash Receipts Assumptions	
3	The annual sales you enter on this row are allocated to months in the cash budget (Part 4) based on the sales pattern you enter (rows 4-9).
4-10	You enter the sales pattern on these rows as monthly percentages. The total (cell F10) should always equal 100% after you enter your figures and recalculate.
12-17	You enter the expected collection pattern on these rows as percentages. You can also enter a trade discount for accounts receivables that are collected within 10 days (cell E13). Total cash receipts (cell F17) should equal 100% after you enter your figures and recalculate.
19	You enter the expected increase in sales for the first quarter of the next year over the first quarter of the year covered by this budget. This estimate is needed so that "leads" can be calculated. For example, inventory purchases and payments are made in the year covered by this budget for sales that occur in the first quarter of the next year.
Part 2. Purchases & Cash Payments Assumptions	
23	You enter the cost of goods sold as a percentage of net sales on this row.
24	You enter the number of months in advance that inventory is purchased on this row. This value is used on row 85 to calculate when inventory purchases are made relative to net sales on row 65.
26-30	You enter the payment pattern for purchases on these rows as percentages. These percentages are used by formulas on rows 88-91 to calculate the months in which cash payments are made. The total (row 30) should equal 100% after you enter your figures and recalculate.
32	The variable operating expense percentage you enter here is used by formulas on row 94 to calculate monthly variable operating expenses.
33	The monthly fixed operating expenses you enter here are carried to row 95.
34	Quarterly taxes you enter here are carried to row 96.
38-39	You enter the advertising expense pattern here. It's used to calculate monthly advertising expenses (row 97) by assuming that expenditures are made equally over the quarter.
Part 3. Short-Term Financing Assumptions	
46	The minimum cash balance you enter here establishes the minimum cash balance. If the cash balance on row 110 falls below the minimum, borrowing is calculated (row 114) to bring the cash balance up to the minimum required balance.
47	The opening cash balance you enter here is carried to cell C 106.
48	The money market interest rate you enter here is used to calculate interest earned (row 107).
52	The opening notes payable balance you enter here is carried to cell C 113.
53	The notes payable interest rate you enter here is used to calculate interest paid (row 108).
57	The opening accounts receivable balance you enter here is used to calculate the first month's cash receipts on rows 70-72.
58	The opening accounts payable balance you enter here is used to calculate the first month's cash payments on rows 90 and 91.
Part 4. Sales & Cash Receipts Budget	
63	The dates are carried from Part 1.
65	Monthly net sales are calculated by multiplying the monthly sales percentage you entered on rows 4-9 times the annual sales you entered on row 3.

Table 6 (Cont.)

68-72	Cash receipts are calculated by multiplying the percentages you entered (rows 12-16) times the monthly net sales (row 65). Receipts collected within the discount period (row 69) are discounted by the trade discount you entered (cell E13). The first month's cash receipts on rows 70-72 are based on the opening accounts receivables you entered (row 57).
73-77	Blank rows are provided in case you want to explore the impact of additional cash receipts.
79	Formulas on this row calculate total monthly cash receipts.
Part 5. Purchases & Cash Payments Budget	
83	The dates are carried from Part 1.
85	Total purchases are calculated by multiplying the cost of goods sold percentage you entered (row 23) times monthly sales. Monthly purchases "lead" monthly sales by the number of months that you indicated on row 24.
88-91	Cash payments are calculated by multiplying the percentages you entered (rows 26-29) times the monthly total purchases (row 85). The first month's cash payments on rows 90 and 91 are based on the opening accounts payables you entered (row 58).
93	Formulas on this row sum the monthly cash payments calculated on rows 88-91.
94	Variable operating expenses are calculated by multiplying the percentage you entered (row 32) times monthly net sales (row 65).
95	Fixed monthly operating expenses are carried from row 33.
96	Quarterly tax payments are carried from row 34.
97	Advertising expenses are carried from rows 38 and 39.
99	Total cash payments are the sum of rows 93-97.
Part 6. Cash Budget Summary	
105	Net monthly cash flow is total cash receipts (row 79) minus total cash payments (row 99).
106	Opening cash/money market balance in month one is carried from row 47. For subsequent months, it is the previous month's ending balance (row 119).
107	Interest earned is the opening cash balance (row 106) times the interest rate (row 48) divided by 12.
108	Interest paid is the opening notes payable balance (row 113) times the interest rate (row 53) divided by 12.
110	Cash available adds net monthly cash flow (row 105), opening cash/money market balance (row 106), and interest earned (row 107) minus interest paid (row 108).
113	Opening notes payable is carried from row 52 in month one and from the previous month's closing notes payable balance (row 117) for subsequent months.
114	Borrowing is calculated when the cash available (row 110) falls below the minimum cash balance (row 46). The amount calculated is the desired cash balance minus the actual balance before borrowing.
115	Notes repaid are calculated if the cash available (row 110) is above the minimum desired balance (row 46). This is limited by the opening notes payable balance (row 113).
117	Closing notes payable balance is the sum of opening notes payable (row 113) and borrowing (row 114) minus the notes repaid (row 115).
119	Ending cash/marketable securities is cash available (row 110) and borrowing (row 114) minus notes repaid (row 115).

THE TEMPLATE'S GRAPHS

The CASHFLOW template has two bar graphs that you can display if your computer is equipped to display graphics.

Graph 1. Cash Receipts and Disbursements plots the monthly total cash receipts (row 79) and total cash payments (row 99) over the period of the budget (row 121).

Graph 2. Closing Notes Payable plots the monthly closing notes payable (row 117) over the period of the budget (row 121).

CASE 1 LEISURE WEAR — PART I

Leisure Wear, a chain of 15 women's clothing stores, is developing the cash budget for 19x2. Total sales are expected to be $8,500,000, with the following monthly pattern:

MONTH	%	MONTH	%	MONTH	%
January	6	May	8	September	9
February	6	June	7	October	9
March	7	July	7	November	10
April	8	August	8	December	15

- Currently, 80% of the sales are for cash. The 20% that are credit sales are collected within 30 days, and no discount is offered. Sales for the first quarter of 19x3 are expected to be 10% higher than the first quarter of 19x2.
- Cost of goods sold averages 52% of sales price. Goods are purchased one month before they are sold. Payment is made within 30 days with no discount.
- Variable operating expenses average 18% of sales. Fixed operating expenses are $170,000 per month. Tax payments are $27,000 per quarter. In addition, advertising is planned to take advantage of the prime sale periods. The advertising plan is as follows:

QUARTER	AMOUNT	QUARTER	AMOUNT
First	$40,000	Third	$45,000
Second	$60,000	Fourth	$60,000

Leisure Wear deposits all its cash in a money market account earning 6% interest. It likes to maintain a minimum balance in this account of $50,000, which is also the opening balance. Other relevant January 1, 19x2 balances are accounts receivable, $383,000; accounts payable, $265,000; and notes payable, $135,000. The interest rate on notes payable is 10%.

REQUIREMENTS

A. Prepare the cash budget for 19x2. What is the maximum closing notes payable balance? Which months require short-term borrowing? Why is there so little short-term borrowing in this seasonal business?

B. To meet the competition, we decide to extend more credit. This results in a drop in our cash sales to 70%, with 25% of our sales collected within 30 days and the remaining 5% collected within 60 days. How does this affect the maximum closing notes payable balance and the months that require borrowing? Explain.

C. Suppose a 2% cash discount for payment within 10 days is offered. This would alter the collection pattern in Requirement B so that 20% would pay within the discount period, 5% within 30 days, and 5% within 60 days. Contrast the results of this discount on the closing notes payable balance and the months that require borrowing with the results of Requirement B.

D. Return to the Requirement B customer credit assumptions. If 25% of our suppliers allowed 60 days to pay for purchases, what would be the maximum closing notes payable balance and the months that require borrowing?

E. From this analysis, summarize the impact of customer and supplier credit on the cash needs of a seasonal business.

Name: ______________________

CASE 1. LEISURE WEAR — PART I SOLUTION

Template: CASHFLOW

DATA ANALYSIS SUMMARY (attach model printouts as supporting documents):

Leisure Wear
Cash Budget Summary

REQUIREMENT	MAXIMUM NOTES PAYABLE	MONTHS BORROWING REQUIRED
A		
B		
C		
D		

DISCUSSION

REQUIREMENTS

A. Prepare a cash budget for 19x2, the first year of operation. How large does the notes payable balance get and can it be repaid in 19x2?

B. Prepare a cash budget for 19x2 assuming revenues are only $150,000 but everything else remains the same. How large does the notes payable balance get and can it be repaid in 19x2?

C. Several small, professional offices have discovered that your rates are lower than commercial cleaning services. If they contract for your service, they want 30-day credit terms (no discount). If 40% of your projected $225,000 in revenue is granted 30-day credit terms, how does the maximum notes payable balance and repayment schedule differ from Requirement A? Assume all credit customers pay within 30 days.

D. Return to the assumption that the $225,000 in projected revenue is all immediate cash receipts. Suppose you decide to hire a salaried supervisor for the independent contractors. This will decrease your variable operating expenses to 40% of revenue but increase your fixed operating expenses to $10,000 per month. Compare the resulting maximum notes payable balance and repayment schedule with that calculated in Requirement A. What impact does this change in policy have on your cash needs?

E. From this analysis, summarize the short-term financing issues of a startup business.

Name: ______________________________

CASE 3. SPIC & SPAN CLEANING SERVICE SOLUTION

Template: CASHFLOW

DATA ANALYSIS SUMMARY (attach model printouts as supporting documents):

Spic & Span Cleaning Service
Cash Budget Summary

REQUIREMENT	MAXIMUM BORROWING		MONTH LOAN REPAID
	AMOUNT	MONTH	
A			
B			
C			
D			

DISCUSSION

TUTORIAL 5 USING FIXED TITLES AND WINDOWS

The CASHFLOW template is both long and wide. This makes it difficult to understand the relationships within the template unless you can bring distant parts together on the screen so that you can see them at the same time. You can do this with the Titles and Window commands.

- Fixed titles keep selected rows and columns from scrolling off the screen when you move the cell pointer around the worksheet.
- Windows split the screen into two vertical or horizontal windows through which you can display different parts of the template. These parts can then be scrolled together (synchronized) or independently (unsynchronized).

Let's see how these commands work.

Step 1 Retrieve CASHFLOW. If the CASHFLOW template is not on the screen, retrieve it from the disk on which you saved it.

Step 2 Page Down to Part 4 and Fix Titles. The Titles command fixes rows above the cell pointer and columns to its left. Begin by moving to Part 4. Align the screen so that cell A61 is in the upper left corner of the screen, and then move the cell pointer to cell C67.

To fix titles

Press **/WTB** (for *Worksheet*, *Titles*, *Both*)

Result. No change is obvious on the screen, but press **Tab** three times. The columns scroll so that you can see additional months, but column A remains fixed on the screen. Press **Home** to move the cell pointer to the upper-left corner of the unfixed range. Now press **PgDn** repeatedly. The rows scroll, but rows 61 through 66 remain fixed on the screen. Press **Home** to move the cell pointer to the upper-left corner of the unfixed range.

Step 3 Clear the Fixed Titles. You can remove fixed titles at any time.

To clear the fixed titles

Press **/WTC** (for *Worksheet*, *Titles*, *Clear*)

Result. You can now move freely around the template.

Step 4 Fix Column A. On this template, and many others, you want only to keep a column from scrolling off the screen. Begin by displaying Part 4 on the screen, and then move the cell pointer to column C. (Column B contains the vertical line that you will also fix on the screen.)

To fix column A

Press **/WTV** (for *Worksheet*, *Titles*, *Vertical*)

Result. Press **PgDn** and then **Tab** and **Shift-Tab**. You can now move freely around the template while the row labels in column A always remain on the screen.

To clear the fixed titles

Press **/WTC** (for *Worksheet*, *Titles*, *Clear*)

Step 5 Create a Horizontal Window. A horizontal window splits the screen so that the row in which the cell pointer is positioned becomes the topmost row in the lower window.

To create a horizontal window

Press **Home** to display all of Part 1 on the screen
Goto A11, press **/WWH** (for *Worksheet*, *Window*, *Horizontal*)

Result. The screen splits into two windows with the cell pointer in the top window. Press **Window (F6** on the IBM PC) repeatedly, and watch the cell pointer jump back and forth between windows. Move the cell pointer within one of the windows, and then press **Window (F6)** two more times. See how the cell pointer always jumps back to the same cell it was in when it last left the window.

Press **Window (F6)** to move the cell pointer into the lower window. Press **Goto**, type **PART 4** and then press **Enter** to move to Part 4. Press → to scroll the screen at least until column G comes into view. See how both parts scroll together. This is because the windows are synchronized. Press **Window (F6)** to move the cell pointer into the upper window, and then press **Home** to align Part 1 on the screen. Press **/WWU** (for *Worksheet*, *Window*, *Unsync*). Press **Window (F6)** to move the cell pointer into the lower window. Now press → again, and you can scroll the lower window without scrolling the upper.

Step 6 Clear the Window. You can return the display to a single window at any time.

To clear the window

Press **/WWC** (for *Worksheet*, *Window*, *Clear*)

EXERCISE 6
LEVEL 0-3 VARIANCE ANALYSIS

In this exercise, you solve cases that are based on the principles of variance analysis discussed in Chapters 7 and 8 of *Cost Accounting: A Managerial Emphasis* (8th edition). This exercise assumes that the template VARIANCE is displayed on the screen. (If you need help on retrieving files, refer to Exercise 1 or Appendix C.)

UNDERSTANDING THE TEMPLATE

The VARIANCE template has nine parts (Figure 27 and Table 7). Parts 1 through 6 calculate the Level 0 to Level 3 variance analysis described in Chapter 7 of *Cost Accounting: A Managerial Emphasis* (8th edition) and illustrated in Exhibit 7-9. Part 7 integrates standard product costing into Level 3 of the analysis by calculating direct material and direct labor (Chapter 7) and variable and fixed overhead (Chapter 8) standard cost variances.

Parts 8 and 9 already contain the input variables for the Allen Company, which is the basis for the cases in this exercise. The case requirements ask you to change some of these input variables in Parts 8 and 9 and observe the impact on the Level 0 to Level 3 variance analysis calculated by Parts 1 through 7.

Part 1. Analysis Level 0 calculates Level 0 variance based on the input variables in Parts 8 and 9.

Part 2. Analysis Level 1 calculates Level 1 variance S based on the input variables in Parts 8 and 9.

Part 3. Analysis Level 2 calculates Level 2 variance S based on the input variables in Parts 8 and 9.

Part 4. Analysis Level 2only — **Detailed Variable Cost Variances** calculates detailed Level 2 variable cost variances based on the input variables in Parts 8 and 9.

Part 5. Analysis Level 2 — Detailed Fixed Cost Variances calculates detailed Level 2 fixed cost variances based on the input variables in Parts 8 and 9.

Part 6. Analysis Level 3 — Sales Price Variance by Territory calculates sales price variances by territory based on the input variables in Parts 8 and 9.

Part 7. Analysis Level 3 — Standard Product Cost Variances calculates standard product cost variances based on the input variables in Parts 8 and 9.

Part 8. Input Variables — Sales contains the sales input variables.

Part 9. Input Variables — Costs contains the cost input variables.

FIGURE 27 The VARIANCE Template.

```
          A           B        C          D           E           F          G
 1 PART 1. ANALYSIS LEVEL 0         Brackets () indicate unfavorable variances
 2 ============ ======= ========= ========= =========== ========= ===========
 3
 4
 5
 6   Actual operating income         $709,469       87.4%
 7   Budgeted operating income        812,200      100.0%
 8                                   ---------- -----------
 9   Static budget variance          (102,731)     -12.65%
10                                   ========= ===========
11
12
13
14
15
16
17
18
19
20 ============ ======= ========= ========= =========== ========= ===========
21 PART 2. ANALYSIS LEVEL 1
22 ============ ======= ========= ========= =========== ========= ===========
23                                                Static
24                                               (Master
25                         Actual                 budget)              Variance
26 ------------ ------- --------- --------- ----------- --------- -----------
27 Units                   20,015          %      19,800        %         215
28                      --------- --------- ----------- --------- -----------
29 Revenues (sales)     7,536,575    100.0% $7,524,000    100.0%      12,575
30 Variable costs       3,095,106     41.1%  2,989,800     39.7%   (105,306)
31                      --------- --------- ----------- --------- -----------
32 Contribution margin  4,441,469     58.9%  4,534,200     60.3%    (92,731)
33 Fixed costs          3,732,000     49.5%  3,722,000     49.5%    (10,000)
34                      --------- --------- ----------- --------- -----------
35 Operating income      $709,469      9.4%   $812,200     10.8%   (102,731)
36                      ========= ========= =========== ========= ===========
37                          |___________________|
38                                    |
39 Static budget variance          (102,731)
40 ============ ======= ========= ========= =========== ========= ===========
41 PART 3. ANALYSIS LEVEL 2
42 ============ ======= ========= ========= =========== ========= ===========
43                                 Flexible                  Sales      Static
44                                   budget    Flexible     volume     (Master
45                         Actual  variances     budget  variances     budget)
46 ------------ ------- --------- --------- ----------- --------- -----------
47 Physical units/output   20,015          0      20,015       215      19,800
48                      --------- --------- ----------- --------- -----------
49 Revenues (sales)     7,536,575   (69,125)  7,605,700    81,700   7,524,000
50 Variable costs       3,095,106   (72,841)  3,022,265  (32,465)   2,989,800
51                      --------- --------- ----------- --------- -----------
52 Contribution margin  4,441,469  (141,966)  4,583,435    49,235   4,534,200
53 Fixed costs          3,732,000   (10,000)  3,722,000         0   3,722,000
54                      --------- --------- ----------- --------- -----------
55 Operating income      $709,469  (151,966)   $861,435    49,235    $812,200
56                      ========= ========= =========== ========= ===========
57 Flex. budgt. variances   |____  (151,966) ____|
58 Sales volume variances                        |_____   49,235   ____|
59 Static budgt. variance   |____  _________  ($102,731) _________ ____|
60 ============ ======= ========= ========= =========== ========= ===========
```

FIGURE 27 (Cont.)

	A	B	C	D	E	F	G
61	PART 4. ANALYSIS LEVEL 2 - DETAILED VARIABLE COST VARIANCES						
63					Flexible		
64					budget		Flexible
65			Actual		variances		budget
67	Number of units		20,015		0		20,015
69	[VARIABLE COSTS]						
71	Direct materials		$889,776		(9,116)		$880,660
72	Direct labor		1,048,110		(47,360)		1,000,750
73	Factory Overhead		304,220		(3,995)		300,225
75	Mfg. cost		2,242,106		(60,471)		2,181,635
76	Selling and Admin		853,000		(12,370)		840,630
78	Total variable costs		3,095,106		(72,841)		3,022,265
81	PART 5. ANALYSIS LEVEL 2 - DETAILED FIXED COST VARIANCES						
83					Flexible		
84					budget		Flexible
85			Actual		variances		budget
87	Number of units		20,015		0		20,015
89	[FIXED COSTS]						
91	Mfg.overhead		1,967,000		13,000		1,980,000
92	Selling and Admin		1,765,000		(23,000)		1,742,000
94	Total fixed costs		3,732,000		(10,000)		3,722,000
101	PART 6. ANALYSIS LEVEL 3 - SALES PRICE VARIANCE BY TERRITORY						
105		Sales			Sales		
106		territory			price		
108		1			($22,625)		
109		2			26,700		
110		3			(73,200)		
111		4			0		
113		Total			($69,125)		

CASE 2 ALLEN COMPANY — PART II

The Allen Company manufactures a single electronic measuring device on which it holds the U.S. patent rights. Because this device has quality control applications in many industries and it is significantly more effective than the competition, Allen Company has been very successful in marketing it nationally.

For cost control purposes, Allen Company has installed a budgeting system that is supported by a standard cost system within its manufacturing facility. Another factor that contributes to the Company's success is its state-of-the-art production planning system. Through techniques such as just-in-time delivery of materials and production scheduling driven by a computerized customer order system, the company has been able to:

1. Virtually eliminate finished goods inventory (the units produced equal the units sold) and
2. Virtually eliminate raw materials inventory (material price variance can be isolated when the material is used).

The actual performance for the month of April, 19x1, along with the budget and standard cost information for the Allen Company, has been entered into the template. This case examines the direct material and direct labor standard cost variances as part of the Level 0 to Level 3 variance analysis described in Chapter 7 of your text and illustrated in Exhibit 7-5.

Each requirement is independent.

REQUIREMENTS

A. Analyze the performance of the Allen Company for the month of April, 19x1. To accomplish this, first prepare a schedule that demonstrates how the lower-level variances can be combined to form higher-level variances.

B. Assume that Allen Company actually paid $998,200 for direct labor. (Enter this change in Part 9 of the template in the Actual Results section.) How would this change affect the flexible budget (Level 2), total direct labor (Level 2), and direct labor rate (Level 3) variances? Explain.

C. Assume that Allen Company had not been so efficient in its use of direct material, and actual direct material units used were 40,030 at a cost of $892,899. (Enter this change in Part 9 of the template in the Actual Results section.) How would this change affect the flexible budget (Level 2), total direct material (Level 2), and direct material usage (Level 3) variances? Explain.

Name:________________________

CASE 2. ALLEN COMPANY — PART II SOLUTION
Template: VARIANCE

DATA ANALYSIS SUMMARY (attach model printouts as supporting documents):

Allen Company
Variance Analysis – Direct Material & Labor
For the Month Ended April 30, 19x1

Requirement	Flexible Budget	Direct Labor		Direct Material	
		Total	Rate	Total	Usage
A	($151,966)	($47,360)	($49,910)	($9,116)	$3,080
B				NR	NR
C		NR	NR		

NR = not required.

DISCUSSION

CASE 3 ALLEN COMPANY – PART III

The Allen Company manufactures a single electronic measuring device on which it holds the U.S. patent rights. Because this device has quality control applications in many industries and it is significantly more effective than the competition, Allen Company has been very successful in marketing it nationally.

For cost control purposes, Allen Company has installed a budgeting system that is supported by a standard cost system within its manufacturing facility. Another factor that contributes to the Company's success is its state-of-the-art production planning system. Through techniques such as just-in-time delivery of materials and production scheduling driven by a computerized customer order system, the Company has been able to:

1. Virtually eliminate finished goods inventory (the units produced equal the units sold) and
2. Virtually eliminate raw materials inventory (material price variance can be isolated when the material is used).

The actual performance for the month of April, 19x1, along with the budget and standard cost information for the Allen Company, has been entered into the template. This case examines the variable and fixed manufacturing overhead variances discussed in Chapter 8 of your text. It integrates these variances with the Level 0 to Level 3 variance analysis described in Chapter 7 and illustrated in Exhibit 7-5.

Each requirement is independent.

REQUIREMENTS

A. Analyze the performance of the Allen Company for the month of April, 19x1. To accomplish this, first prepare a schedule that demonstrates how the lower-level variances can be combined to form higher-level variances.

B. Assume the following changes in Allen Company's performance: (1) actual direct labor hours, 100,075; (2) actual variable overhead, $298,500; and (3) actual direct labor cost, $1,050,787. (Enter these changes in Part 9 of the template in the Actual Results section.) How would these changes affect the flexible budget (Level 2) and variable overhead spending and efficiency (Level 3) variances? Explain.

C. Assume the following changes in Allen Company's performance: (1) actual direct labor hours, 100,075; (2) actual fixed manufacturing overhead, $1,960,000; and (3) actual direct labor cost, $1,050,787. (Enter these changes in Part 9 of the template in the Actual Results section.) How would these changes affect the flexible budget (Level 2) and fixed manufacturing overhead budget and volume (Level 3) variances? Explain.

D. Summarize the impact of actual direct labor hours and actual manufacturing overhead costs on the computation of manufacturing overhead variances.

Name:_

CASE 3. ALLEN COMPANY — PART III SOLUT
Template: VARIANCE

DATA ANALYSIS SUMMARY (attach model printout

Allen Company
Variance Analysis — Manufacturing
For the Month Ended April 30,

REQUIREMENT	FLEXIBLE BUDGET	VARIABLE SPENDING	VARIABL EFFICIENC
A	($151,966)	($4,760)	$765
B			
C		NR	NR

NR = not required.

DISCUSSION

TUTORIAL 6 PROTECTION AND RECALCULATION

In this tutorial, you explore protection and recalculation.

- You protect cells so that data cannot be entered into them. As you have seen, entering data into a cell deletes any data that was previously there. If you enter a number or label in a cell containing a formula, the formula is deleted. Since this can cause serious damage to a template, you protect key cells so that their contents can't be damaged.
- Recalculation takes place after you enter data. All formulas are recalculated so that they display the correct result. When set to automatic recalculation, the entire template recalculates whenever you enter a number. On small templates, this just takes a second, so you don't notice any delays. On large templates, however, recalculation can take a long time. In those cases, you can set recalculation to manual so that formulas are recalculated only when you press **Calc (F9** on the IBM PC).

Step 1 Retrieve the VARIANCE Template. Retrieve the template if it isn't already on the screen.

Step 2 Explore Global Protection. Begin by moving to Part 8, where you input variables. Try entering a number in cell C143, the cell containing the label *Quantity*. The computer beeps.

To turn protection off

Press **/WGPD** (for *Worksheet, Global, Protection, Disable)*

Result. No results are seen on the screen, but try entering the number again. This time you can enter it because you have turned protection off.

Step 3 Explore Range Protection. Your program has two kinds of protection commands. The first kind is a global command; it turns protection on and off. When on, you can't enter data into any of the spreadsheet's cells. When off, you can enter data anywhere. Protecting all the cells is occasionally helpful, but normally you want to enter data into selected cells. To do this, you use the second kind of protection command, the range protect and unprotect command. Here, turn global protection back on so that you can't enter data into cell C143. Then unprotect just that cell.

To turn protection on

Press **/WGPE** (for *Worksheet, Global, Protection, Enable)*

To unprotect a cell

Goto C143, press **/RU** (for *Range, Unprotect*) and then press **Enter**

Result. With the cell pointer in cell C143, a *U* appears on the status line to remind you it is an unprotected cell. Try entering a number again. You can enter it even though global protection is on because you have turned protection off for the cell. Type **"Quantity** in the cell to restore the label. Then press **/RP** (for *Range*, *Protect*), and then press **Enter** to protect the cell again.

Step 4 Explore Recalculation. The VARIANCE template's recalculation is set to automatic. When you enter data, it recalculates automatically. Here, set recalculation to manual.

To change recalculation

Press **/WGR** (for *Worksheet*, *Global*, *Recalculation*)
Press **M** (for *Manual*)

Result. Enter a number in one of the input cells. See the *CALC* indicator appear on the screen. This indicates that data has been entered but the formulas have not been recalculated. The template may be displaying incorrect answers in the cells containing formulas. Press **Calc** (**F9** on the IBM PC), and watch the indicator disappear. The formulas have now all recalculated.

Step 5 Check the Status of Protection and Recalculation. You can always check your current settings with the Worksheet Status command. Press any key to return to the template.

To check your worksheets status

Press **/WS** (for *Worksheet*, *Status*)

Result. A screen appears indicating the current settings for these and other global commands. Press any key to return to the template.

EXERCISE 7
COST ESTIMATION WITH REGRESSION ANALYSIS

In this exercise, you solve cases that are based on the principles of simple least squares regression analysis discussed in Chapter 10 of *Cost Accounting: A Managerial Emphasis* (8th edition). This exercise assumes that the template COSTBEHV is displayed on the screen. (If you need help on retrieving files, refer to Exercise 1 or Appendix C).

UNDERSTANDING THE TEMPLATE

The COSTBEHV template has three parts (Figure 28 and Table 8). You select a case from the template menu, and then select the independent and dependent variables in Part 1. Using this information, Part 2 calculates the simple least squares regression equation and the coefficient of determination, and it provides a place for you to enter a projected volume for cost estimation. Part 3 contains the data points used in the calculations.

The data for the cases in this exercise are already stored on the disk, so you do not have to enter it from the keyboard. To load data for a case into the template, select *Data/Cases* from the template menu, and select the appropriate case number.

The COSTBEHV template is a large worksheet, so recalculation takes a long time. For this reason, recalculation has been set to manual. The *CALC* indicator in the lower right corner of the screen is on when the worksheet has not been recalculated. After making choices from the menu, or entering your own data, press **Calc** (**F9** on the IBM PC) to recalculate the worksheet.

The *CIRC* indicator on the bottom row of the screen indicates that the template contains a circular reference. Circular references occur when formulas refer to themselves either directly or indirectly. The circular reference in this template is a result of the template's design and does not affect the accuracy of the output.

Part 1. Menu of Variables displays the menu for the case you select from the template menu. From this menu, you can select the data to be used as the dependent and independent variables.

Part 2. Computations calculates the simple least squares regression equation, coefficient of determination, and t-value of the regression coefficient, and it provides a place for you to enter a projected volume for cost estimation.

Part 3A. Data contains the data points used in the calculations.

FIGURE 28 The COSTBEHV Template.

```
        A          B            C          D          E            F
 1 PART 1. MENU OF VARIABLES
 2 =========== ============= =========== =========== ============= ===========
 3
 4                              VARIABLE CHOICES
 5
 6                              [1] Choice 1
 7                              [2] Choice 2
 8                              [3] Choice 3
 9                              [4] Choice 4
10
11             After selecting a case from menu, enter numbers
12             for variable choices from above list and press
13             CALC (F9 on the IBM PC). The mode indicator in
14             the lower-right or upper-right corner reads
15             "WAIT" while the model is recalculating.
16
17             Enter independent variable ------>               [   ]
18             Enter dependent variable -------->               [   ]
19
20 =========== ============= =========== =========== ============= ===========
21 PART 2. COMPUTATIONS
22 =========== ============= =========== =========== ============= ===========
23
24 # data points                                   N           NA
25 Intercept                                       a           NA
26 Slope coefficient                               b           NA
27 Coefficient of determination                  r^2           NA
28 t-value of the regression coefficient           t           NA
29
30                         REGRESSION EQUATION
31                         ----------- -------
32                    Y'=        NA          +                 NA  X
33
34
35             Input projected volume                       #______
36
37             Projected variable cost                          NA
38             Projected total cost                             NA
39
40 =========== ============= =========== =========== ============= ===========
```

FIGURE 28 (Cont.)

	A	B	C	D	E	F
41	PART 3A. DATA					
42	===========	=============	===========	===========	=============	===========
43	Data		Independent		Dependent	
44	point		variable		variable	
45			(x)		(y)	
46	-----------	-------------	-----------	-----------	-------------	-----------
47	NA		NA		NA	
48	NA		NA		NA	
49	NA		NA		NA	
50	NA		NA		NA	
51	NA		NA		NA	
52	NA		NA		NA	
53	NA		NA		NA	
54	NA		NA		NA	
55	NA		NA		NA	
56	NA		NA		NA	
57	NA		NA		NA	
58	NA		NA		NA	
59	NA		NA		NA	
60	NA		NA		NA	
61	NA		NA		NA	
62	NA		NA		NA	
63	NA		NA		NA	
64	NA		NA		NA	
65	NA		NA		NA	
66	NA		NA		NA	
67	NA		NA		NA	
68	NA		NA		NA	
69	NA		NA		NA	
70	NA		NA		NA	
71	NA		NA		NA	
72	NA		NA		NA	
73	NA		NA		NA	
74	NA		NA		NA	
75	NA		NA		NA	
76	NA		NA		NA	
77	NA		NA		NA	
78	NA		NA		NA	
79	NA		NA		NA	
80	===========	=============	===========	===========	=============	===========

Table 8
THE COSTBEHV TEMPLATE

Rows	Description
Part 1. Menu of Variables	
6-9	These rows list the choices available for the independent and dependent variables. You select any two of these choices on the rows described below.
17-18	Enter the number of one of the variables listed on rows 6-9 above for an independent and for a dependent variable.
Part 2. Computations	
24	The number of data points in the current analysis is displayed on this row.
25	The intercept is calculated.
26	The slope coefficient is calculated.
27	The coefficient of determination is calculated.
28	The t-value of the regression coefficient is calculated.
32	The regression equation formula is displayed along with the current values.
35	You enter the projected volume on this row.
37	The projected variable costs is the slope coefficient (row 26) times the projected volume (row 35).
38	The projected total cost is the intercept or fixed costs (row 25) plus the projected variable costs (row 37).
Part 3A. Data	
A	The number of data points is displayed in this column. If a row displays *NA*, that row has no data in columns C or E.
C	The independent variable you selected on row 17 of Part 1 is carried from Part 3B of the template, where all the case data is stored. Press **Tab** to see the case data section; then press **Shift-Tab** to move back to this section.
E	The dependent variable you selected on row 18 of Part 1 is carried from Part 3B of the template, where all the case data is stored.

THE TEMPLATE'S GRAPH

The COSTBEHV template has a scatter graph that you can display if your computer is equipped to display graphics. This graph also displays the simple least squares regression line calculated in Part 2.

Graph 1. Plot of Regression Analysis plots the dependent variable (column E) from Part 3A against the independent variable (column C) from Part 3A. Each data point is plotted as a square. The simple least squares regression line is plotted as a solid line.

CASE 1 MERCY HOSPITAL

Before beginning the preparation of next year's budget, Rick Smart, the controller of Mercy Hospital, accumulates data on the costs of operating each department over the last several years and the related volume of hospital activity. The typical measure of hospital activity is patient days. The data for a ten-year period for three departments is reproduced below (000 omitted):

YEAR	PATIENT DAYS	X-RAY	PHARMACY	OTHER
1	252	$1,881	$2,314	$817
2	253	1,882	2,358	857
3	255	1,886	2,402	854
4	256	1,884	2,445	856
5	257	1,880	2,480	847
6	261	1,887	2,431	862
7	268	1,888	2,529	866
8	273	1,896	2,626	877
9	278	1,898	2,557	881
10	280	1,896	2,730	896

REQUIREMENTS

A. Using simple least squares regression, determine the flexible budget formula for each department.

B. What do these flexible budget formulas indicate about the cost structure of each department? How good an estimate is each of these formulas?

C. If the hospital forecasts 277 thousand patient days for next year, prepare the budget for each department from your flexible budget formula. What are the significant limitations of these budgets?

Name:________________________

CASE 1. MERCY HOSPITAL SOLUTION
Template: COSTBEHV

DATA ANALYSIS SUMMARY (attach model printouts as supporting documents):

Mercy Hospital
Cost Estimation & Projection

DEPARTMENT	FIXED COST	VARIABLE RATE/PD	t	r2	BUDGET FOR 277,000 PD
X-ray					
Pharmacy					
Other					

DISCUSSION

CASE 2 NORR MANUFACTURING

For product costing, budgeting, and control, Norr Manufacturing needs to estimate the behavior of its manufacturing overhead. Susan Hansen, the budget analyst, is assigned the task of making this estimate. Since she is uncertain about the best measure of manufacturing volume, she gathers data on several measures of volume as well as the related manufacturing overhead costs for the last 12 periods. This information (000 omitted) is reproduced below:

PERIOD	MANUFACTURING OVERHEAD	DIRECT LABOR HOURS	MACHINE HOURS	UNITS PRODUCED
1	$484	42	19	21
2	506	35	20	19
3	600	50	25	23
4	433	30	17	18
5	613	49	26	26
6	690	52	30	26
7	508	41	20	22
8	679	52	29	25
9	614	45	26	27
10	637	51	27	28
11	727	55	32	28
12	541	46	32	23

REQUIREMENTS

A. Using simple least squares regression, determine the formula for estimating the manufacturing overhead cost using each of the three volume measures. Which volume measure appears best?

B. Examine the scatter diagrams of each volume measure. What do you observe from this? What may have caused this?

C. Based on your observations in Requirement B, rerun the regression estimate for the volume measure in question. (Data points may be changed or eliminated in Part 3B. To move to this part, press **Goto** (**F5** on the IBM PC), type **PART 3B** and then press **Enter**.) Does this change your recommendation about the best volume measure?

Name:________________________

CASE 2. NORR MANUFACTURING SOLUTION

Template: COSTBEHV

DATA ANALYSIS SUMMARY (attach model printouts as supporting documents):

Norr Manufacturing
Cost Estimation — Manufacturing Overhead

VOLUME MEASURE	FIXED COST	VARIABLE RATE	t	r2
Direct labor hours				
Machine hours				
Units produced				
- Revised				

DISCUSSION

CASE 3 HIGH TECH COMPANY

Rapid growth has been very difficult to manage for the officers of High Tech Company. They are quite concerned that costs are getting out of control because everyone in the company is emphasizing meeting customer demand. Manufacturing overhead is selected for the initial analysis because of its significance. Data (000 omitted) for the first 24 months of operation using direct labor hours as a measure of volume is presented below.

MONTH	DIRECT LABOR HOURS	MANUFACTURING OVERHEAD
1	10	$64.5
2	12	67.5
3	12	69.5
4	13	70.5
5	14	70
6	15	71
7	15	74
8	16	72
9	17	73
10	17	75
11	18	75
12	18	76.5
13	19	78
14	19	79
15	20	80
16	20	82
17	20	84
18	21	85.5
19	22	86.5
20	22	89.5
21	23	90
22	23	93
23	24	93
24	24	99

REQUIREMENTS

A. Using all 24 months of data, compute the equation for the simple least squares regression line. Examine the scatter diagram of these data points. What do you observe?

B. Assume that your investigation of the data for month 24 indicates that manufacturing overhead costs increased because of a supplier's strike. Recompute the simple least squares regression line using only the first 23 data points. (Data points may be changed or eliminated in Part 3B. To move to this part, press **Goto**, (**F5** on the IBM PC), type **PART 3B** and then press **Enter**.) Explain why this adjustment in the analysis is being made.

C1. Using the first 12 months of data, recompute the equation for the simple least squares regression line.

C2. Using the data for months 13 to 23, recompute the equation for the simple least squares regression line.

Compare these two equations, and speculate on why they differ.

D. Assume that you discover that there was a major price increase in month 13. Removing this price increase from the manufacturing overhead costs would result in the following:

MONTH	REVISED MFG. OH	MONTH	REVISED MFG. OH
13	77	19	85.4
14	78	20	88.4
15	79	21	88.8
16	81	22	91.8
17	83	23	91.8
18	84.4		

Recompute the equation for the simple least squares regression line for months 13 to 23. How does this affect your discussion of Requirement C?

Name:__________________________

CASE 3. HIGH TECH COMPANY SOLUTION

Template: COSTBEHV

DATA ANALYSIS SUMMARY (attach model printouts as supporting documents):

High Tech Company
Manufacturing Overhead Estimation

REQUIREMENT	FIXED COST	VARIABLE RATE/DLH	t	r2
A				
B				
C1				
C2				
D				

DISCUSSION

TUTORIAL 7 ENTERING YOUR OWN DATA

The COSTBEHV template is a powerful tool that can easily be adapted to your own data set. To use it to analyze another data set, just clear the template and enter the data in Part 3B. If you want to reuse the data at a later time, however, you must extract it into its own file on the disk, and then revise the menu of variable choices and the template menu to add it as a new case. Let's see how you would add Case 4.

Step 1 Retrieve the COSTBEHV Template. Retrieve the template if it isn't on the screen. To turn off protection so that you can work on the template, press **/WGPD**.

Step 2 Clear Data from the Template. To adapt the template for your own data, clear any existing data from the template.

To clear data

Press	**Alt-M** to display the template menu
Press	**DC** (for *Data/Cases, Clear*)
Press	**Enter** when the warning is displayed

Result. The data is removed from the template, and the cell pointer is positioned in the upper-left corner of Part 3B. The mode indicator reads *WAIT* while the template is recalculated.

Step 3 Enter and Save Your Data. Press **Q** (for *Quit-menu*) to remove the menu. Move the cell pointer to the columns in which you want to enter numbers, and then type them in. Once your data is entered, you can extract it into its own file on the disk.

To extract the new data

Press	**/FXV** (for *File, Xtract, Values*), and the prompt asks you to enter the name for the file in which data will be saved
Type	**CASE4** and then press **Enter**. The prompt asks you to enter the range to be extracted
Type	**PART 3B** and then press **Enter**

Result. The drive spins as your data are saved into the file named CASE4 on the disk.

Step 4 Revise the Menu. You saw how macros worked in Exercise 1. In this step and the following, you revise the macro that automatically changes the menu in Part 1 of the template and retrieves your data. You begin by copying the menu that appears when you select the Clear command and then add the new case to the menu.

To go to the clear menu

Press	**Goto** (**F5** on the IBM PC), and the prompt asks you to enter the cell address you want to go to
Type	**AB41** and then press **Enter**

To copy the clear menu

Press	**/C** (for *Copy*), and the prompt asks you to enter the range to copy FROM
Type	`AB47.AB52` and then press **Enter**. The prompt asks you to enter the range to copy TO
Type	`AC47` and then press **Enter**
Press	→ to see the copied menu

To revise the menu

Goto	AC45, type `Case 4` and then press **Enter**
Goto	AC47, and then press **Edit** (F2 on the IBM PC) to display *VARIABLE CHOICES* on the edit line
Press	**Spacebar**, type `- CASE 4` and then press **Enter**

To name the new menu

Press	**/RNC** (for *Range, Name, Create*), and the prompt asks you to enter the range's name
Type	`CASE4` and then press **Enter**. The prompt asks you to enter the range being named
Type	`AC47.AC52` and then press **Enter**

Step 5 Revise Macro. Now that you have entered and saved your data and created a new menu of variables, you can revise the macro to add the new case to the menu.

To move the clear macro

Goto	AC26, press **/M** (for *Move*), and the prompt asks you to enter the range to move FROM
Press	↓ to expand the highlight down to row 33, and then press **Enter**. The prompt asks you to enter the range to move TO
Press	→ to point to cell AD26, and then press **Enter**

Result. The clear menu is moved from column AC to column AD.

To copy the macro for Case 3

Goto	AB26, press **/C** (for *Copy*), and the prompt asks you to enter the range to copy FROM
Press	↓ to expand the highlight down to row 32, and then press **Enter**. The prompt asks you to enter the range to copy TO
Press	→ to point to cell AC26, and then press **Enter**
Press	→ to see the copied macro

To revise the macro for Case 4

Goto	AC26, type `'4` and then press **Enter**. (The ' enters the 4 as a label.)
Goto	AC27, type a description of your case, and then press **Enter**
Goto	AC29, and then press **Edit** to display *'{goto}CASEDATA~/fcceCASE3~* on the edit line
Press	**Backspace** twice to delete the *3~* in *CASE3~*

Type	**4~** and then press **Enter** so that the macro in cell AC29 reads *{goto}CASEDATA~/fcceCASE4~*
Goto	AC31, and then press **Edit**
Press	← to position the edit cursor under *3* in *CASE3,* and then press **Del**
Type	**4** and then press **Enter** so that the macro in cell AC31 reads */cCASE4~MENUS~*

FIGURE 29 The Completed Macro. After entering your macro, your screen should look like this figure.

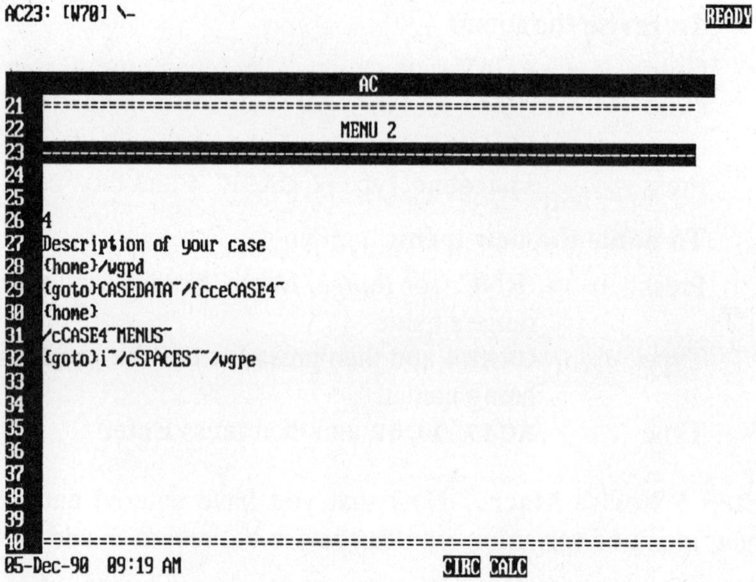

Result. Your macro should look like Figure 29. Hold down **Alt-M** to display the template menu. Press **D** (for *Data/Cases*). The new case is listed on the menu following Case 3. Before executing the macro, back up and save the file. This saves your work in case anything goes wrong. If you really get fouled up, copy the file COSTBEHV.WKS from your original *Cost Accounting Resource Disk* to your backup disk, and then start over or quit. (Note that each version of the 1-2-3 program has its own unique extension. Lotus 1-2-3 Release 1 uses .WKS, and Lotus 1-2-3 Releases 2.01 and 2.2 use .WK1. After copying the file, rename it so that it has the correct extension.)

EXERCISE 8
COST ALLOCATION: SERVICE DEPARTMENT CHARGES

In this exercise, you solve cases that are based on the principles of allocating service department charges discussed in Chapter 14 of *Cost Accounting: A Managerial Emphasis* (8th edition). This exercise assumes that the template SVCDEPT is displayed on the screen. (If you need help on retrieving files, refer to Exercise 1 or Appendix C.)

UNDERSTANDING THE TEMPLATE

The SVCDEPT template has two parts (Figure 30 and Table 9). You enter data into Parts 1A and 1B, and Part 2 calculates the service department allocations to the user departments automatically.

Parts 1A and 1B. Departmental Service Usage and Service Department Costs are where you enter budgeted and actual departmental usage and service department costs.

Part 2. User Department Allocations calculates the service department allocations based on the variables you enter in Part 1.

FIGURE 30 The SVCDEPT Template.

```
              A              B         C         D         E         F         G
 1  PART 1A. DEPARTMENTAL SERVICE USAGE
 2  =============== ========= ========= ========= ========= ========= =========
 3  User depts.             1         2         3         4         5     Total
 4  --------------- --------- --------- --------- --------- --------- ---------
 5  Usage-Budget    #_______  #_______  #_______  #_______  #_______         NA
 6  Usage-Actual    #_______  #_______  #_______  #_______  #_______         NA
 7
 8  =============== ========= ========= ========= ========= ========= =========
 9  PART 1B. SERVICE DEPARTMENT COSTS             () = unfavorable variance
10  =============== ========= ========= ========= ========= ========= =========
11  Actual usage           NA
12
13                                                          [--- Variance ---]
14  Costs                          Rate    Actual    Budget        $         %
15  --------------- --------- --------- --------- --------- --------- ---------
16  Variable costs            $_______  $_______         NA        NA        NA
17  Fixed costs                   ----  $_______  $_______         NA        NA
18  --------------- --------- --------- --------- --------- --------- ---------
19  Total                                     NA        NA        NA        NA
20  =============== ========= ========= ========= ========= ========= =========
21  PART 2. USER DEPARTMENT ALLOCATIONS
22  =============== ========= ========= ========= ========= ========= =========
23  User depts.             1         2         3         4         5     Total
24  --------------- --------- --------- --------- --------- --------- ---------
25  Actual usage           NA        NA        NA        NA        NA        NA
26  =============== ========= ========= ========= ========= ========= =========
27  Service Cost Allocation
28
29  Var. costs             NA        NA        NA        NA        NA        NA
30  Fix. costs             NA        NA        NA        NA        NA        NA
31  --------------- --------- --------- --------- --------- --------- ---------
32  Total                  NA        NA        NA        NA        NA        NA
33  =============== ========= ========= ========= ========= ========= =========
34
35                                       -- NOTES --
36
37  (1) Variable costs are allocated based upon actual service usage.
38  (2) Fixed costs are allocated based upon budgeted service usage.
39
40  =============== ========= ========= ========= ========= ========= =========
```

Table 9
THE SVCDEPT TEMPLATE

Rows	Description
Part 1A. Departmental Service Usage	
5	Budgeted service usage is entered on this row for up to five departments.
6	Actual service usage is entered on this row for up to five departments.
Part 1B. Service Department Costs	
11	Actual usage is the total actual usage carried from cell G6.
16	The variable cost rate and actual variable costs are entered on this row. Budgeted variable costs are the variable cost rate (cell C16) times total actual usage (cell G6). Variance in dollars is the actual variable costs (cell D16) minus the budgeted variable costs (E16). Brackets indicate unfavorable variances. Variance as a percentage (cell G16) is the variance in dollars (cell F16) divided by the budgeted variable costs (cell E16).
17	The actual and budgeted fixed costs are entered on this row. Variance in dollars (cell F17) is the actual fixed costs (cell D17) minus the budgeted fixed costs (E17). Brackets indicate unfavorable variances. Variance as a percentage (cell G17) is the variance in dollars (cell F17) divided by the budgeted fixed costs (cell E17).
19	Total costs and total variances are the sum of variable costs (row 16) and fixed costs (row 17).
Part 2. User Department Allocations	
25	Actual usage by department is carried from Part 1 (row 6).
29	Allocated variable costs are the department's actual usage (row 6) times the variable cost rate (cell C16).
30	Allocated fixed costs are the department's budgeted usage (row 5) divided by total budgeted usage (cell G5). The resulting percentage is then multiplied times budgeted fixed costs (cell E17).
32	The total departmental allocation is the sum of the variable and fixed cost allocations.

CASE 1 STATE PRINTING SERVICE

The numerous agencies of state government make such extensive use of outside printing services that a State Printing Service (SPS) was created. By having such a centralized printing service, each agency would have access to high-quality but lower-priced printing services. The high quality would be available because the volume of printing from all agencies justified the purchase of sophisticated printing equipment and maintenance of a professional staff. No single agency could justify such an expenditure. The low price could be achieved partly because of the efficiency of the printing equipment and partly because the printing service was to be operated on a breakeven basis.

The fixed costs of the SPS, such as equipment leases and salaries, are budgeted at $80,000 per month. These costs will be allocated based on each agency's budgeted hours of printing usage. The budgeted variable costs of operating the SPS are $25 per hour of usage. These costs will be allocated based on each agency's actual hours of printing service usage each month.

Assume there are five state agencies that will be billed for SPS usage. The relevant data for the month of September, 19x1, is shown below.

USER DEPARTMENT	1	2	3	4	5
Budgeted hours*	75	150	50	250	100
Actual hours	80	150	60	210	85

*Long-run average usage.

During this month, the SPS incurred $15,950 of variable costs and $82,000 of fixed costs.

REQUIREMENTS

A. Compute the SPS cost allocation for each state agency for the month.

B. If the fixed costs of operating SPS were allocated based on actual usage, compute the cost allocation to each state agency for the month. (*Hint:* Insert the actual usage into the budgeted usage row.)

C. From the perspective of the agencies (users), discuss the motivational aspects of allocating fixed costs based on budgeted volume as compared to actual volume. Use the previous two requirements as an illustration.

D. Review the SPS departmental cost report. What can be concluded about the performance of the SPS? Explain how allocations based on standard or budgeted prices assist in the analysis.

CASE 3 SWANSON STORES

Swanson Stores is a department store chain that assigns each store manager broad responsibilities for store operation. One function that is centralized is credit and collection because Swanson Stores has its own charge card, which is the only form of credit accepted by all the stores.

The measure for the volume of activity in the credit and collection department is the number of credit transactions. More complicated transactions, such as performing a credit check for a new account or collection procedures for late payments are assigned weights so that all activity can be reduced to the number of transactions measure.

The fixed costs of operating the credit and collection department are $30,000 per month. These costs will be allocated based on each store's budgeted number of credit transactions. The budgeted variable costs of the credit and collection department are $0.15 per credit transaction. These costs will be allocated based on each store's actual number of credit transactions each month.

The relevant data for the five-store chain for the month of April, 19x1, is reproduced below.

USER STORE	1	2	3	4	5
Budgeted transactions*	10,000	25,000	14,500	8,000	19,000
Actual transactions	10,872	24,350	12,480	8,960	18,120

*Long-run average usage.

During the month, the credit and collection department incurred $11,345 of variable costs and $28,000 of fixed costs.

REQUIREMENTS

A. Compute the credit and collection department cost allocation for each store for the month.

B. Assume Swanson Stores decides to measure the volume of activity in the credit and collection department with total dollars of credit transactions. The relevant information for the five stores for the month of April, 19x1, is (000 omitted):

USER STORE	1	2	3	4	5
Budgeted $ *	$500	$890	$630	$400	$760
Actual $	525	875	560	450	725

*Long-run average usage.

Recompute the credit and collection department cost allocation for each store for the month. Use dollars of credit transactions to measure the volume of activity and $3.60 per thousand dollars of credit transactions as the budgeted variable cost rate.

C. Which volume measure, number of credit transactions or dollars of credit transactions, is the better measure for allocating the cost of the credit and collection department? Relate your discussion to the previous two requirements.

Name:____________________________

CASE 3. SWANSON STORES SOLUTION
Template: SVCDEPT

DATA ANALYSIS SUMMARY (attach model printouts as supporting documents):

Swanson Stores
Credit & Collection Department Allocations
For the Month Ended April 30, 19x1

	User Stores				
Requirement	1	2	3	4	5
A					
B					

DISCUSSION

TUTORIAL 8 EXTRACTING AND COMBINING FILES

If you are working in a company that has a number of divisions or that updates reports on a regular basis, you can use your program's file combine commands to combine data stored in separate files.

Step 1 Retrieve the SVCDEPT template. Retrieve the template, and then enter data used in one of your cases.

Step 2 Extract the Data into Its Own Disk File. Let's begin by extracting the data on rows 5 and 6 and save them in a file on the disk.

To extract a range of numbers

Goto	B5, press **/FXV** (for *File*, *Xtract*, *Values*), and the prompt asks you to enter the name of the file in which the extracted data will be saved
Type	**USAGE** and then press **Enter**. The prompt asks you to enter the range to be extracted
Press	→ and ↓ to expand the lower-right corner of the highlight to cell F6, and then press **Enter**

Result. The drive spins as the extracted range is copied to the file named USAGE on the disk.

Step 3 Erase Existing Data. Now that the numbers have been extracted, you can erase them from the template.

To erase the existing data

Goto	B5, press **/RE** (for *Range*, *Erase*), and the prompt asks you to enter the range to be erased
Press	→ and ↓ to expand the lower-right corner of the highlight to cell F6, and then press **Enter**

Result. The range is erased.

Step 4 Copy the File. Now, use the Combine command to copy the numbers back into the template. The cell pointer's position is important because the file you are combining from the disk will appear with its upper-left corner in the cell in which you position the cell pointer.

To combine the files

Goto	B5, press **/FCCE** (for *File*, *Combine*, *Copy*, *Entire File*), and the prompt asks you to enter the name of the file to be combined
Type	**USAGE** and then press **Enter**

Result. The numbers you extracted reappear in the template.

Step 5 Add the File. You just used the Copy command, which reads in numbers. If other numbers are in the cells into which they are copied, they overwrite and erase them. The Combine command also has a choice that lets you add the numbers that are combined to the numbers already in the cells.

To add numbers

Goto	B5, press **/FCAE** (for *File*, *Combine*, *Add*, *Entire File*), and the prompt asks you to enter the name of the file to be combined
Type	**USAGE** and then press **Enter**

Result. The original numbers you stored in the USAGE file are added to the numbers in the range on the template into which they are combined.

Step 6 Subtract the File. You can also subtract the combined numbers from the numbers already in the range.

To subtract numbers

Goto	B5, press **/FCSE** (for *File*, *Combine*, *Subtract*, *Entire File*), and the prompt asks you to enter the name of the file to be combined
Type	**USAGE** and then press **Enter**

Result. The original numbers you stored in the USAGE file are subtracted from the numbers in the range on the template.

EXERCISE 9
COST ALLOCATION: PRODUCT OVERHEAD COST

In this exercise, you solve cases that are based on the principles of allocating product overhead cost discussed in Chapter 15 of *Cost Accounting: A Managerial Emphasis* (8th edition). This exercise assumes that the template PRODCOST is displayed on the screen. (If you need help on retrieving files, refer to Exercise 1 or Appendix C.)

Note: If you are not using an Epson or Epson-compatible printer, and you did not enter setup codes for your printer in Exercise 1 or in another exercise, you have to enter them before you can print this template. You can do so when the PRODCOST template is displayed on the screen. To do so, refer to the box "Tip on Entering Printer Setup Codes" in Exercise 5, on page 83.

UNDERSTANDING THE TEMPLATE

The PRODCOST template has five parts (Figure 31 and Table 10). Data have been entered into the template, which illustrates the determination of overhead rates for product costing.

Parts 2 through 5 are wider than the screen, so you have to scroll sideways to see all of them. Press **Tab** to scroll to the right one screen. Hold down **Shift** and press **Tab** to move back to the left one screen. To keep the row labels in column A from scrolling off the screen when you press **Tab**, see Appendix I, "Using Fixed Titles."

Part 1. Input Screen is the only part of the template in which you enter data for the cases.

Part 2. Service Department and Production Department Costs contains all the direct departmental cost data.

Part 3. Service Department Allocations performs the service department allocations and the calculation of production department predetermined overhead rates from the selections you made in Part 1 and the departmental cost data in Part 2.

Part 4. Overhead Assignment — Sample Jobs contains information about two sample jobs. These sample jobs were added to the data to illustrate the effect of different factory overhead rates on the amount of factory overhead applied to different jobs. You may substitute information on other sample jobs on these rows to determine the impact of different predetermined overhead rates.

Part 5. Allocation Bases contains the service department (expressed as percentages) and production department (expressed in quantities) allocation bases used in these calculations. You may substitute other allocation bases on these rows to determine their impact on the predetermined overhead rates.

FIGURE 31 The PRODCOST Template.

	A	B	C	D	E	F	G	H	I
1	PART 1. INPUT SCREEN								
2	======================	=========	===========	==========	===========				
3	Service department - Base selection								
4									
5	Department	Allocation Bases			Selection				
6	Repair	1=Specific service 2=Equipment $			1				
7	Buildings & grounds	1=Square footage 2=Service time			1				
8	Factory administration	1=DLH 2=# of employees			1				
9	----------------------	---------	-----------	----------	-----------				
10	Service department - Allocation method								
11		Method			Selection				
12		1=Step 2=Direct			1				
13	----------------------	---------	-----------	----------	-----------				
14	Overhead application bases								
15									
16	Department	Application Bases			Selection				
17	Machining	1=DLH 2=Machine hours			1				
18	Assembly	1=DLH 2=DL $			1				
19	Finishing	1=DLH 2=DL $			1				
20	======================	=========	===========	==========	===========				
21	PART 2. SERVICE DEPARTMENT COSTS					PRODUCTION DEPARTMENT COSTS			
22	======================	=========	===========	==========	===========	===========	===========	===========	===========
23	LINE		Repair and	Buildings &	Factory				
24			maintenance	grounds	admin	Machining	Assembly	Finishing	Total
25	----------------------	---------	-----------	----------	-----------	-----------	-----------	-----------	-----------
26	1. Indirect labor		$400,000	$960,000	$2,200,000	$4,900,000	$1,000,000	$360,000	$9,820,000
27	2. Supplies		640,000	700,000	240,000	3,500,000	400,000	1,400,000	6,880,000
28	3. Power		130,000	1,600,000	25,000	1,800,000	200,000	500,000	4,255,000
29	4. Payroll taxes		200,000	50,000	200,000	1,300,000	300,000	100,000	2,150,000
30	5. Overtime premium		100,000	----	60,000	600,000	140,000	50,000	950,000
31	6. Rework		----	----	----	300,000	150,000	40,000	490,000
32	7. Fuel		----	100,000	----	----	----	----	100,000
33	----------------------	---------	-----------	----------	-----------	-----------	-----------	-----------	-----------
34	8. Total variable		1,470,000	3,410,000	2,725,000	12,400,000	2,190,000	2,450,000	24,645,000
35	9. Property tax		140,000	770,000	760,000	600,000	100,000	100,000	2,470,000
36	10. Depreciation		150,000	700,000	300,000	2,300,000	400,000	300,000	4,150,000
37	11. Supervision		700,000	220,000	4,015,000	1,800,000	1,010,000	550,000	8,295,000
38	----------------------	---------	-----------	----------	-----------	-----------	-----------	-----------	-----------
39	Totals		2,460,000	5,100,000	7,800,000	17,100,000	3,700,000	3,400,000	39,560,000
40	======================	=========	===========	==========	===========	===========	===========	===========	===========
41	PART 3. SERVICE DEPARTMENT ALLOCATIONS								
42	======================	=========	===========	==========	===========	===========	===========	===========	===========
43	Allocation method:		Repair &	Buildings &	Factory				
44	If 1=Step 2=Direct ->	1	maintenance	grounds	admin	Machining	Assembly	Finishing	Total
45	----------------------	---------	-----------	----------	-----------	-----------	-----------	-----------	-----------
46									
47	12. Total department costs		2,460,000	5,100,000	7,800,000	17,100,000	3,700,000	3,400,000	39,560,000
48									
49	13. Repair - allocation		(2,460,000)	492,000	24,600	1,476,000	246,000	221,400	2,460,000
50				-----------					
51	14. Buildings & grounds - allocation			5,592,000	559,200	3,355,200	1,118,400	559,200	5,592,000
52				==========	-----------				
53	15. Factory admin OH - allocation				8,383,800	6,707,040	838,380	838,380	8,383,800
54					==========	-----------	-----------	-----------	-----------
55	16. Total production dept OH					28,638,240	5,902,780	5,018,980	39,560,000
56						==========	==========	==========	==========
57	17. Application bases					2,400,000	300,000	300,000	
58									
59	18. Predetermined overhead rate					$11.933	$19.676	$16.730	
60	======================	=========	===========	==========	===========	===========	===========	===========	===========

FIGURE 31 (Cont.)

	A	B	C	D	E	F	G	H	I
61	PART 4. OVERHEAD ASSIGNMENT - SAMPLE JOBS								
62									
63		Overhead	[------ Job 101	------	[------ Job 102	------]			
64		rate	Usage	$	Usage	$			
65									
66	Direct material	----		$5,000		$8,000			
67	Direct labor	----		4,000		12,000			
68									
69	Factory overhead	----							
70	Machining - DLH	$11.93	300	3,580	25	298			
71	- Mach hour	$0.00	150	0	10	0			
72	Assembly - DLH	$19.68	25	492	300	5,903			
73	- DL$	$0.00	$125	0	$1,600	0			
74	Finishing - DLH	$16.73	25	418	25	418			
75	- DL$	$0.00	$475	0	$375	0			
76				-----------		-----------			
77	Total factory overhead			$4,490		$6,619			
78				===========		===========			
79	Total costs			$13,490		$26,619			
80									
81	PART 5. ALLOCATION BASES								
82									
83	LINE		Repair &	Buildings &	Factory				
84			maintenance	grounds	admin	Machining	Assembly	finishing	Total
85									
86	Service Department								
87	1. Specific services		----	20.0%	1.0%	60.0%	10.0%	9.0%	100.0%
88	2. Equipment $		----	15.0%	0.5%	75.5%	5.0%	4.0%	100.0%
89	3. Square footage		----	----	10.0%	60.0%	20.0%	10.0%	100.0%
90	4. Service time - B&G		----	----	5.0%	45.0%	30.0%	20.0%	100.0%
91	5. Direct labor hours (DLH)		----	----	----	80.0%	10.0%	10.0%	100.0%
92	6. # of employees		----	----	----	55.0%	20.0%	25.0%	100.0%
93									
94	Overhead Application								
95									
96	1. Direct labor hours (DLH)		----	----	----	2,400,000	300,000	300,000	
97	2. Machine hours		----	----	----	1,000,000	NA	NA	
98	3. Direct labor dollars (DL$)		----	----	----	NA	$1,500,000	$4,500,000	
99									
100									

Table 10
THE PRODCOST TEMPLATE

Rows	Description
Part 1. Input Screen	
6-8	You select one of the two bases for allocating each service department's costs on these rows when solving Case 1.
12	You select either the step or direct allocation method on this row when solving Case 2.
17-19	You select one of the two bases for applying each production department's overhead to products when solving Case 3.
Part 2. Service/Production Department Costs	
26-32	These rows contain the variable costs for three service and three production departments.
34	Total variable costs are the sum of the variable costs entered on rows 26-32.
35-37	These rows contain the fixed costs for three service and three production departments.
39	Total costs for each department are the sum of all costs on rows 34-37.
Part 3. Service Department Allocations	
44	The service department allocation method you selected in Part 1 (row 12) is carried here (cell B44).
47	Total department costs are carried to this row from Part 2 (row 39).
49-53	Costs are allocated from each service department to the production departments based on your entries in Part 1.
55	Total production department overhead is the sum of the department costs and the service department allocations (rows 49-53).
57	If you enter 1 on rows 17-19, the application bases from row 96 are carried here. If you enter 2 on rows 17-19, the application bases from row 97 or 98 are carried here.
59	Predetermined overhead rates are the total production department overhead (row 55) divided by the application bases (row 57).
Part 4. Overhead Assignment – Sample Jobs	
66	The direct material for the two sample jobs.
67	The direct labor for the two sample jobs.
70-75	Factory overhead is calculated on these rows (see below).
77	Total factory overhead is the sum of the overhead items on rows 70-75.
79	Total costs are the sum of total factory overhead (row 77), direct material (row 66), and direct labor (row 67).
B	Predetermined overhead rates are carried from row 59. Which of the two rates for each department are used depends on the overhead application base you specified in Part 1 on rows 17-19.
C & E	Usage is entered in these columns for each of the two sample jobs.
D & F	Dollars of factory overhead for each job are the overhead rate (column B) times usage (columns C and E).
Part 5. Allocation Bases	
87-92	Service department allocation base percentages are entered on these rows for the three service departments.
96-98	Overhead application bases are entered on these rows for the three production departments.

CASE 1 SAMPLE COMPANY – PART I

Chapter 15 of your text discusses allocation of the indirect costs of production to determine applied overhead rates for product costing. Data from the Sample Company have already been entered in the template to illustrate this process. Two hypothetical production jobs have also been included in the template to illustrate the impact of the applied overhead rates on the amount of overhead assigned to different jobs.

This case specifically examines the impact of selecting a base or cost driver to allocate service department costs on the applied overhead rates. To do this, a second possible allocation base for each service department has been added to the example in developing the template.

REQUIREMENTS

A. Compute the applied overhead rate for each of the three production departments if equipment dollars is used as the base to allocate the repair and maintenance department. Describe the reason that the overhead rates changed. What effect did this change have in the amount of overhead applied to Job 101 as compared to Job 102? Why?

B. Return the repair and maintenance department to its specific service allocation base. Compute the applied overhead rate for each of the three production departments if service time is used as the base to allocate the building and grounds department. Describe the reason that the overhead rates changed. What effect did this change have on the amount of overhead applied to Job 101 as compared to Job 102? Why?

C. Return the building and grounds department to its square footage allocation base. Compute the applied overhead rate for each of the three production departments if number of employees is used as the base to allocate the factory administration department. Describe the reason that the overhead rates changed. What effect did this change have on the amount of overhead applied to Job 101 as compared to Job 102? Why?

Name:____________________________

CASE 1. SAMPLE COMPANY — PART I SOLUTION
Template: PRODCOST

DATA ANALYSIS SUMMARY (attach model printouts as supporting documents):

Sample Company
Service Allocation Bases and Overhead Rates

	OVERHEAD RATES/DLH			TOTAL O.H. APPLIED	
REQUIREMENT	**MACHINING**	**ASSEMBLY**	**FINISHING**	**JOB 101**	**JOB 102**
Example	$11.933	$19.676	$16.730	$4,490	$6,619
A					
B					
C					

DISCUSSION

CASE 2 SAMPLE COMPANY — PART II

Chapter 15 of your text discusses allocation of the indirect costs of production to determine applied overhead rates for product costing. Data from the Sample Company have already been entered in the template to illustrate this process. Two hypothetical production jobs have also been included in the template to illustrate the impact of the applied overhead rates on the amount of overhead assigned to different jobs.

This case specifically examines the impact of selecting the step or direct method of service department allocation on the applied overhead rates.

REQUIREMENTS

A. Compute the applied overhead rate for each of three production departments using the step method of service department allocation. How much overhead is applied to Job 101 and Job 102 using this method?

B. Compute the applied overhead rate for each of the three production departments using the direct method of service department allocation. What effect did this change have on the amount of overhead applied to Job 101 as compared to Job 102? Why?

C. Would the order in which the service departments are allocated affect the applied overhead rates if the direct method were used? The step method?

Name:____________________________

CASE 2. SAMPLE COMPANY — PART II SOLUTION
Template: PRODCOST

DATA ANALYSIS SUMMARY (attach model printouts as supporting documents):

Sample Company
Allocation Method and Overhead Rates

	OVERHEAD RATES/DLH			TOTAL O.H. APPLIED	
REQUIREMENT	MACHINING	ASSEMBLY	FINISHING	JOB 101	JOB 102
A (step)					
B (direct)					

DISCUSSION

CASE 3 SAMPLE COMPANY — PART III

Chapter 15 of your text discusses allocation of the indirect costs of production to determine applied overhead rates for product costing. Data from the Sample Company have already been entered in the template to illustrate this process. Two hypothetical production jobs have also been included in the template to illustrate the impact of the applied overhead rates on the amount of overhead assigned to different jobs.

This case specifically examines the impact of selecting an application base or cost driver for the applied overhead rates. To do this, a second possible application base for each production department has been added to the example in developing the template.

REQUIREMENTS

A. Compute the applied overhead rate for the machining department if machine hours is used as the application base. What effect did this change have on the amount of overhead applied to Job 101 as compared to Job 102? Why?

B. Return the machining department to its direct labor hour application base. Compute the applied overhead rate for the assembly department if direct labor dollars is used as the application base. What effect did this change have on the amount of overhead applied to Job 101 as compared to Job 102? Why?

C. Return the assembly department to its direct labor hour application base. Compute the applied overhead rate for the finishing department if direct labor dollars is used as the application base. What effect did this change have on the amount of overhead applied to Job 101 as compared to Job 102? Why?

Name:______________________________

CASE 3. SAMPLE COMPANY — PART III SOLUTION
Template: PRODCOST

DATA ANALYSIS SUMMARY (attach model printouts as supporting documents):

Sample Company
Application Bases and Overhead Rates

	OVERHEAD RATES			TOTAL O.H. APPLIED	
REQUIREMENT	MACHINING	ASSEMBLY	FINISHING	JOB 101	JOB 102
example	$11.933/DLH	$19.676/DLH	$16.730/DLH	$4,490	$6,619
A		19.676/DLH	16.730/DLH		
B	11.933/DLH		16.730/DLH		
C	11.933/DLH	19.676/DLH			

DISCUSSION

TUTORIAL 9 USING DATE FUNCTIONS

Now that you have made it this far you have become quite accomplished with 1-2-3. Let's take a break and explore date functions, one of your program's more interesting features.

Step 1 Erase the Screen. If a template is displayed on the screen, remove it with the worksheet erase command. This command erases anything that is on the screen, so be sure to save your work before you execute it.

To erase the worksheet

Press **/WEY** (for *Worksheet*, *Erase*, *Yes*)

Result. The screen shows the blank worksheet.

Step 2 Enter and Format Dates. Your program has a number of date functions. The interesting thing about these functions is that their dates can be added and subtracted.

To enter a date

Goto A1, type **`@DATE(92,1,1)`** and then press **Enter**

Result. The number 33604 is displayed in the cell. This is the number of days since December 31, 1899. Not much help yet is it?

To format the date

Goto A1, press **/RFD** (for *Range*, *Format*, *Date*), and the prompt reads *1 (DD-MMM-YY) 2 (DD-MMM) 3 (MMM-YY)*

Press **1** and the prompt asks you to enter the range to format

Press **Enter**

Result. The column is filled with asterisks. This indicates the column isn't wide enough for the number and format you have entered. To widen the column, press **/WCS** (for *Worksheet*, *Column*, *Set*). Press → to widen the column until the date is displayed, and then press **Enter**. The date is displayed as 01-Jan-92. Now let's enter a second date in cell A2.

To enter and format a date

Goto A2, type **`@DATE(92,6,20)`** and then press **Enter**

Goto A2, press **/RFD** (for *Range*, *Format*, *Date*), and the prompt reads *1 (DD-MMM-YY) 2 (DD-MMM) 3 (MMM-YY)*

Press **1** (for *(DD-MMM-YY)*), and the prompt asks you to enter the range to format

Press **Enter**

Result. The date 20-Jun-92 is displayed.

Step 3 Manipulate Dates. Dates entered with functions are always values, the number of days since December 31, 1899. Since they are values, you can add and subtract them. Here, enter a formula that calculates the number of days between these two dates.

To enter a formula

Goto	A3, type **+A2-A1** and then press **Enter**

Result. There are 171 days between the two dates. To find out how many days you have been alive, enter your birthdate in cell A1 and today's date in cell A2. The format of the date function is simple, @DATE(year,month,day). The year is entered using just the last two digits. The month is entered as its number where January is 1, February is 2, March is 3, and so on. If you were born on April 3, 1970, the function is @DATE(70,4,3).

Step 4 Date Columns. Often you want to date a number of columns, for example, when creating a monthly budget. You can do it by using date functions and formulas.

To date columns

Goto	A5, type **@DATE(92,1,1)** and then press **Enter**
Goto	B5, type +A5+31 and then press **Enter**
Press	**/RFD** (for *Range*, *Format*, *Date*), and the prompt reads *1 (DD-MMM-YY) 2 (DD-MMM) 3 (MMM-YY)*
Press	**3** (for *3 (MMM-YY)*), and the prompt asks you to enter the range to format
Type	A5.B5 and then press **Enter**
Press	**/C** (for *Copy*), and the prompt asks you to enter the range to copy FROM
Press	**Enter**, and the prompt asks you to enter the range to copy TO
Type	C5.G5 and then press **Enter**

Result. Columns A through G are now dated Jan-92 through Jul-92.

- To clear the screen, press **/WEY** (for *Worksheet*, *Erase*, *Yes*).
- To return to AUTO123 to continue with the next exercise, insert the Cost Accounting Resource Disk into drive B (on a floppy disk system) or drive A (on a hard disk system). Then, press **/FR** (for *File*, *Retrieve*), type **B:AUTO123** (on a floppy disk system) or **A:AUTO123** (on a hard disk system), and then press **Enter**.

EXERCISE 10
CAPITAL BUDGETING

In this exercise, you solve cases that are based on the principles of capital budgeting discussed in Chapters 20 and 21 of *Cost Accounting: A Managerial Emphasis* (8th edition). This exercise assumes that the template CAPBUDGT is displayed on the screen. (If you need help on retrieving files, refer to Exercise 1 or Appendix C.)

Note: If you are not using an Epson or Epson-compatible printer, and you did not enter setup codes for your printer in Exercise 1 or in another exercise, you have to enter them before you can print this template. You can do so when the CAPBUDGT template is displayed on the screen. To do so, refer to the box "Tip on Entering Printer Setup Codes" in Exercise 5 on page 83.

UNDERSTANDING THE TEMPLATE

The CAPBUDGT template has four parts (Figure 32 and Table 11). You enter data into Part 1. To speed up the large amount of data entry, recalculation is set to manual. After you enter data, press **Calc** (**F9** on the IBM PC), and Part 2 calculates the project's cash flow. Parts 3 and 4 then calculate several profitability indicators.

Parts 1 through 3 are wider than the screen so you have to scroll sideways to see all of them. Press **Tab** to scroll to the right one screen. Hold down **Shift** and press **Tab** to move back to the left one screen. To keep the row labels in column A from scrolling off the screen when you press **Tab**, see Appendix I, "Using Fixed Titles."

Part 1A. Project Assumptions is where you enter or change variables regarding the investment and specify a depreciation method.

Part 1B. Depreciation Table calculates depreciation based on straight line (SL), double-declining balance for a five year recovery period (DDB-5), and double-declining balance for a seven year recovery period (DDB-7).

Part 2. Project Cash Flows calculates the project cash flows based on the data you enter in Part 1A.

Part 3. Project Evaluation computes several profitability measures that are used to evaluate capital investment projects. The text discusses the advantages of such measures as net present value, IRR, and the payback period. The discounted payback period has been suggested by other authors as a supplemental measure. As the name implies, it is calculated just like the payback period except it uses discounted cash inflows (row 45) rather than project cash flows (row 39).

Part 4. Data Table contains a Data Table 1 you can use to explore a range of assumptions. A detailed description of this type of data table is given in Tutorial 4.

Part 5. Payback Period Worksheets (rows 83-94) contain calculations used by formulas in other parts of the template. You need not refer to them.

FIGURE 32 The CAPBUDGT Template.

```
           A            B       C    D       E       F       G          H          I       J       K        L      M      N
 1 PART 1A. PROJECT ASSUMPTIONS                                                          PART 1B. DEPRECIATION TABLE
 2 ============== =========== = ======== ======== ====== ========= =========           ====== ========= ====== ====== ======
 3 Depreciable Investment                                  $________                      YEAR   STRAIGHT   DDB5   DDB7
 4 Salvage Value                                           $________                                LINE
 5 Working capital investment                              $________                     ------ --------- ------ ------ ------
 6 Project Life (1 to 10 years)                            #________                          1        0      0      0
 7 Discount rate                                            ________%                         2        0      0      0
 8 Corporate Tax Rate                                       ________%                         3        0      0      0
 9 -------------- ----------- - -------- -------- ------ ------------------                   4        0      0      0
10 Select depreciation method (1-SL, 2-DDB5, 3-DDB7)       #________                          5        0      0      0
11 -------------- ----------- - -------- -------- ------ --------- ---------                   6        0      0      0
12 Operating Cash Flows                                                                      7        0      0      0
13   Sales volume in units                                 #________                          8        0      0      0
14   Unit selling price                                    $________                          9        0      0      0
15   Unit variable operating costs                         $________                         10        0      0      0
16   Fixed operating costs                                 $________                     ------ --------- ------ ------ ------
17 -------------- ----------- - -------- -------- ------ --------- ---------           Totals        0      0      0
18   Inflation rate                                         ________%                    ====== ========= ====== ====== ======
19   Applied to (1=Selling price, 2=VOC & FOC, 3=ALL)      #________
20 ============== =========== = ======== ======== ====== ========= =========
21 PART 2. PROJECT CASH FLOWS
22 ============== =========== = ======== ======== ====== ========= ========= ====== ====== ========= ====== ====== ======
23                        Year |     0        1      2         3          4      5      6         7      8      9     10
24 -------------- ----------- - -------- -------- ------ --------- --------- ------ ------ --------- ------ ------ ------
25 Investment outflows:        |
26   Depreciable               |     0      ---    ---       ---        ---    ---    ---       ---    ---    ---    ---
27   Working capital           |     0      ---    ---       ---        ---    ---    ---       ---    ---    ---    ---
28 Operating Cash Flows:       |
29   Revenues                  |   ---        0      0         0          0      0      0         0      0      0      0
30   Variable operating costs  |   ---        0      0         0          0      0      0         0      0      0      0
31   Fixed operating costs     |   ---        0      0         0          0      0      0         0      0      0      0
32 Other Project Cash Flows:   |
33   Salvage value             |   ---        0      0         0          0      0      0         0      0      0      0
34   Working capital recovery  |   ---        0      0         0          0      0      0         0      0      0      0
35   Taxes on operating flows  |   ---        0      0         0          0      0      0         0      0      0      0
36   Taxes on disposal            ---        0      0         0          0      0      0         0      0      0      0
37   Depreciation tax shield   |   ---        0      0         0          0      0      0         0      0      0      0
38 -------------- ----------- - -------- -------- ------ --------- --------- ------ ------ --------- ------ ------ ------
39   Project cash flows        |     0        0      0         0          0      0      0         0      0      0      0
40 ============== =========== = ======== ======== ====== ========= ========= ====== ====== ========= ====== ====== ======
41 PART 3. PROJECT EVALUATION
42 ============== =========== = ======== ======== ====== ========= ========= ====== ====== ========= ====== ====== ======
43                        Year |     0        1      2         3          4      5      6         7      8      9     10
44 -------------- ----------- - -------- -------- ------ --------- --------- ------ ------ --------- ------ ------ ------
45 Present value of cash flows |    NA        0      0         0          0      0      0         0      0      0      0
46 % investment recovered-PV   | -----     0.0%   0.0%      0.0%       0.0%   0.0%   0.0%      0.0%   0.0%   0.0%   0.0%
47 Net present value           |    NA
48 Simple payback period       |    10    Years     NA    Months
49 Discounted payback period   |    10    Years     NA    Months
50 -------------- ----------- - -------- -------- ------ --------- ---------
51 InterNA l Rate of Return    |  Guess      NPV  |      Guess        NPV
52                             | -------- --------  --------- ---------
53      Start guess            |  0.00%        0  |     60.00%          0
54        0.0000%              | 10.00%        0  |     70.00%          0
55                             | 20.00%        0  |     80.00%          0
56       Increment             | 30.00%        0  |     90.00%          0
57        10.0000%             | 40.00%        0  |    100.00%          0
58                             | 50.00%        0  |
59                             |
60 ============== =========== = ======== ======== ====== ========= ========= ====== ====== ========= ====== ====== ======
```

FIGURE 32 (Cont.)

```
           A              B      C     D        E        F      G         H         I        J      K         L       M      N
 61  PART 4. DATA TABLE
 62  ===============  =========== =  ======== ======== ====== ========= =========
 63  INPUT VALUES     +NPV        |    1. Type /WGPD to turn protection off
 64               NA              |  ----------------------------------------------
 65               NA              |    2. Set input range values and press CALC
 66               NA              |       Starting value                #________
 67               NA              |       Increment                     #________
 68               NA              |  ----------------------------------------------
 69               NA              |    3. Type /DT1 and press RETURN
 70               NA              |  ----------------------------------------------
 71               NA              |    4. Enter or point to input cell
 72               NA              |       Sales volume                        H13
 73               NA              |       Selling price                       H14
 74               NA              |       Variable operating costs            H15
 75               NA              |       Fixed operating costs               H16
 76               NA              |  ----------------------------------------------
 77               NA              |    5. Press RETURN and wait until mode
 78               NA              |       indicator changes from WAIT to
 79               NA              |       READY.
 80  ===============  =========== =  ======== ======== ====== ========= =========
 81  PART 5. PAYBACK PERIOD WORKSHEETS
 82  ===============  =========== =  ======== ======== ====== ========= ========= ======= ====== ========= ====== ====== ======
 83  Simple payback period        |
 84          Years                |              0       2       3         4        5       6        7        8       9      10
 85         Months                |            0.0      NA      NA        NA       NA      NA       NA       NA      NA      NA
 86  Discounted payback period    |
 87          Years                |              1       2       3         4        5       6        7        8       9      10
 88         Months                |             NA      NA      NA        NA       NA      NA       NA       NA      NA      NA
 89                               |
 90                               |
 91  Graph axis 100%              |           100%    100%    100%      100%     100%    100%     100%     100%    100%    100%
 92  Project cash flows (39)      |     0       NA      NA      NA        NA       NA      NA       NA       NA      NA      NA
 93  Net present value            |    NA       NA      NA      NA        NA       NA      NA       NA       NA      NA      NA
 94  % investment recovered-PV    | -----       NA      NA      NA        NA       NA      NA       NA       NA      NA      NA
 95                               |
 96                               |
 97                               |
 98                               |
 99                               |
100  ===============  =========== =  ======== ======== ====== ========= ========= ======= ====== ========= ====== ====== ======
```

Table 11
THE CAPBUDGT TEMPLATE

Row or Column	Description
Part 1A. Project Assumptions	
3-6	Assumptions about the project investment are entered on these rows.
7-8	The discount rate and corporate tax rate are entered on these rows.
10	You select the depreciation method on this row.
13-16	Operating cash flow assumptions are entered on these rows.
18	The anticipated inflation rate is entered on this row.
19	You select which of the variables the inflation rate applies to on this row.
Part 1B. Depreciation Table	
K	Straight-line depreciation is the depreciable investment (cell H3) minus the salvage value (cell H4) divided by the project life (cell H6).
L,M	Double-declining balance applies twice the straight-line rate to the declining book value of the investment, ignoring any salvage value. The Modified Accelerated Cost Recovery System (MACRS) of the tax code requires some fixed assets to be depreciated over a five-year recovery period and others over a seven-year period.
Part 2. Project Cash Flows	
26	The depreciable investment is carried from row 3.
27	The working capital investment is carried from row 5.
29	Revenues are sales volume (row 13) times the sales price (row 14). In years 2-10, the revenues are adjusted for inflation if you entered 1 or 3 on row 19.
30	Variable operating costs are sales volume (row 13) times the unit variable operating costs (row 15). In years 2-10, the variable costs are adjusted for inflation if you entered 2 or 3 on row 19.
31	Fixed operating costs are carried from row 16. In years 2-10, the fixed costs are adjusted for inflation if you entered 2 or 3 on row 19.
33	Salvage value is carried from row 4 and displayed in the last year of the project's life entered on row 6.
34	Working capital returned is carried from row 5 and displayed in the last year of the project's life entered on row 6.
35	Taxes are revenues (row 29) minus costs (rows 30 and 31) times the corporate tax rate entered on row 8.
36	Taxes on disposal are assumed assessed in the year of disposal if the DDB5 or DDB7 depreciation method is selected. They are the *Corporate tax rate* (row 8) times the Disposal value (row 4).
37	Depreciation tax shield is the corporate tax rate (row 8) times the depreciation from Part 1B. The column from that part used depends on the depreciation method you entered on row 10.
39	Project cash flows are the sums of all cash flows detailed on rows 26-37.
Part 3. Project Evaluation	
45	Present value of cash flows is the present value of the cash flows on row 39.
46	Percent of investment recovered is the present value of the cash flows to date (row 45) divided by the project cash outflows in period 0 (cell D45).
47	Net present value is the sum of the values on row 45.
48	The simple payback period is calculated on this row.

Table 11 (Cont.)

49	The discounted payback period is calculated on this row.
54-57	To calculate the IRR, you enter a starting value in cell A54 and an increment value in cell A57. These values set the IRR guess range in columns D and G. The NPV is calculated for each value in the guess range. The IRR is a percentage between the last positive and the first negative in the range. To refine the IRR calculation, enter the percentage to the left of the last positive value in the NPV columns (columns E and H) as the starting value. Then reduce the increment value by a decimal place. For example, the starting increment is .1 (10%), so reduce it to .01 (1%).
Part 4. Data Table	
A	The series of values to be run through the input cell you select from the list on the screen. They start at the starting value you enter in cell H66 and increment at the increment value you enter in cell H67.
B	The coordinates of the output cell, net present value (cell D47), head the column where output from the cell will be captured when you run the data table. The reason +*NPV* appears in the cell is that name is assigned to cell D47.

TIP ON DATA TABLE RANGES

Data tables "remember" the table range and input cell. When you press **/DT1** to execute the command, the table range is highlighted. Just press **Enter** to accept it since the table range never changes. The cell pointer then jumps to the input cell that was specified the last time the table was used. If it is the cell you want to use, just press **Enter** to accept it. If you want to specify a different cell, type in its coordinates (or "point" to it with the cell pointer), and then press **Enter**.

THE TEMPLATE'S GRAPH

The CAPBUDGT template has three graphs that you can display if your computer is equipped to display graphics.

Graph 1. Comparison of Cash Flows plots cash flows (row 92) and the present values of cash flows (row 93) against the periods 1-10 (row 43).

Graph 2. Percentage of Investment Recovered (Present Value) plots the percentage of investment recovered (row 46) against the periods (row 43). It also plots a line at 100% on the Y-axis so that you can see when the investment has been recovered.

Graph 3. Graph of Data Table plots the output values (column B) from the net present value cell (cell D47) against the input values (column A).

CASE 1 SPEEDY COPY SERVICE

A replacement decision must be made for one of the copy machines at the Speedy Copy Service. After the initial screening of alternatives, two copy machines remain as contenders – a new model from Xerox and a reconstructed IBM. The relevant information on the two machines follows:

	XEROX	IBM
Installed price	$14,500	$15,500
Maintenance & usage/year	$ 2,300 plus $.01/copy	$ 4,050
Salvage value	$ 1,500	$ 2,500
Life	5 years	5 years

Speedy Copy expects to charge customers $.05 per copy and sell 180,000 copies. But it is uncertain about its service volume estimate. The $.05 per copy it charges customers and the maintenance and usage fees it pays on the copiers are not expected to change significantly over the next five years.

Speedy Copy service requires a minimum pretax rate of return of 16% on its investments as well as a maximum payback period of three years. Since this is a pretax analysis, the corporate tax rate and depreciation method are not relevant.

REQUIREMENTS

A. Compute the net present value and payback period for the Xerox copier. Prepare a data table for volumes from 140,000 copies, in 5,000 copy increments.

B. Compute the net present value and payback period for the IBM copier. Prepare a data table for volumes from 140,000 copies, in 5,000 copy increments.

C. Examine the data tables you prepared for Requirements A and B. At what volume do the two copiers have approximately the same net present value?

D. Which copier would you recommend purchasing? Support your recommendation with the analysis from Requirements A through C.

CASE 2 SOUTHWESTERN PIPELINE COMPANY

Southwestern Pipeline Company will be competing with several other pipeline companies for a contract to service a newly discovered gas field in its service area. The cost of building the connecting pipeline to this gas field is $685,000, and that estimate is very certain. This project will be analyzed over a ten-year life and have no salvage value.

If they win the contract, the company will have exclusive rights to transport gas from the field. The contract specifies the transport fee ($.03 per million cubic feet (MCF) for this project) to be paid to Southwestern and a guaranteed minimum annual volume (4,500,000 MCF). Southwestern Pipeline estimates that the annual volume is most likely to be 5,200,000 MCF. The transport fee cannot be raised during the life of the project.

The operating costs of the pipeline are substantially fixed and estimated to be $7,900 per year. The operating costs will increase annually with the inflation rate. The company estimates that the inflation rate will be 3% annually over the life of the project.

Southwestern expects a pretax rate of return of at least 16%. Since this is a pretax analysis, the corporate tax rate and depreciation method are not relevant.

REQUIREMENTS

A. Compute the net present value for the project using the most likely estimates stated in the case. Does this project meet the required rate of return?

B. Assume that the project will also require an additional investment in working capital of $30,000. Does the project still meet the required rate of return? Explain why the net present value calculated in this requirement differs from that calculated in Requirement A.

C. In addition to the $30,000 investment in working capital, assume that the inflation rate on fixed operating costs will be 8%. Does the project still meet the required rate of return?

D. Continue with the $30,000 investment in working capital and 8% inflation rate assumptions. Prepare a data table for volumes from 4,500,000 MCF, in 100,000 MCF increments. At which of these volumes does the net present value become positive?

E. Would you recommend building this pipeline? Explain.

Name: ______________________

CASE 2. SOUTHWESTERN PIPELINE COMPANY SOLUTION
Template: CAPBUDGT

DATA ANALYSIS SUMMARY (attach model printouts as supporting documents):

Southwestern Pipeline Company
Pipeline Project — Sensitivity Analysis

REQUIREMENT	ANNUAL VOLUME	INFLATION RATE	WORKING CAPITAL	NET PRESENT VALUE
A	5,200,000 MCF	3%	$ 0	
B	5,200,000 MCF	3	30,000	
C	5,200,000 MCF	8	30,000	
D	MCF	8	30,000	

DISCUSSION

CASE 3 GLF PARTNERS

Gary, Larry, and Frank are considering the formation of a partnership to buy and lease equipment. They already have several companies interested in leasing equipment from them. The partnership will invest $120,000 in the equipment, which will have a salvage value of $15,000 at the end of its ten-year life. In addition, the partnership will require an investment of $10,000 in working capital.

GLF expects the partnership to earn a 12% after-tax internal rate of return (IRR) and have an after-tax payback period less than five years. The partnership will not be taxed directly, so the partners will use their marginal personal income tax rates of 31% in the analysis. Revenues will vary with the number of machines leased by the partnership. The plan is to acquire ten machines, which can each be leased for $3,800 annually. Repair costs vary also with the number of machines and are projected to be $100 per machine annually. The fixed costs, such as insurance, are projected to be $12,000 per year. Variable and fixed operating costs are not expected to increase significantly over the life of the project.

REQUIREMENTS

A. Select the straight-line (SL) depreciation option. Does the partnership meet the after-tax IRR and payback requirements? What is the net present value?

B. Select the SL depreciation option. Using the data table in Part 4 of the template (starting value $3,800, increment $25), at what annual lease price per machine does the net present value become positive?

C. Select the double-declining balance seven-year (DDB7) depreciation option. Does the partnership meet the after-tax IRR and payback requirements? What is the net present value?

D. Select the DDB7 depreciation option. Using the data table in Part 4 of the template (starting value $3,800, increment $50), at what annual lease price per machine does the net present value exceed $10,000?

Name:________________________

CASE 3. GLF PARTNERS SOLUTION
Template: CAPBUDGT

DATA ANALYSIS SUMMARY (attach model printouts as supporting documents):

GLF Partners
Equipment Investment – Tax Effects

REQUIREMENT	DEPRECIATION METHOD	LEASE PRICE PER MACHINE	IRR	PAYBACK	NPV
A	SL	$3,800			
B	SL		NR	NR	
C	DDB7	3,800			
D	DDB7		NR	NR	

NR = not required.

DISCUSSION

CASE 4 MIDWEST TELEPHONE COMPANY

Whenever Midwest Telephone Company requests a rate increase, the regulated rate department must prepare the financial data in support of that increase to submit to the State Rate Regulation Board. As manager of the regulated rate department, one of your responsibilities is to monitor the department's extensive use of the mainframe computer.

Midwest Telephone charges all its departments for mainframe computer usage based on their actual computer usage measured in CPU time. Each department receives a budget for computer usage expense each year, and that budget is considered hard dollars. That is, the department manager can spend the department computer budget to buy computer services outside the company.

Every year the computer budget is one of your major headaches. Your budget never is adequate, and you have to request supplements. Further, the mainframe response times are sometimes unsatisfactory when you are under extreme pressure to meet the deadlines imposed by the State Rate Regulation Board for submission of financial information.

Because of these problems, you are exploring the possibility of purchasing a computer system for your department. This expenditure, $150,000, is large enough to require review by the Capital Investment Committee. You decide to use an eight-year project life with no salvage value.

The Capital Investment Committee has set a minimum after-tax internal rate of return (IRR) of 14% for all projects. They also expect a project to have a payback period less than half the project life. Assume a combined federal and state income tax rate of 41%.

An analysis of past computer usage indicates that the regulated rate department uses 2,800 hours of computer time annually, for which it has been internally charged $60 per hour. The variable cost of operating its own computer system is $10 per hour, and the annual fixed operating costs are $80,000. Variable and fixed operating costs are expected to rise at a 5% annual rate.

REQUIREMENTS

A. For each depreciation method listed below, report the resulting after-tax net present value, IRR, and payback period.

A1. Select the straight-line (SL) depreciation option,
A2. Select the double-declining balance 5 year (DDB5) depreciation option,
A3. Select the double-declining balance 7 year (DDB7) depreciation option, and
A4. Select the DDB5 depreciation option and assume a 10% Investment Tax Credit (deduct 10% from the depreciable investment).

What do you conclude about the impact of tax policy on capital budgeting decisions?

B. Select the DDB5 depreciation option. How low can the hourly rate for mainframe charges be while still having the project's net present value remain positive? (*Hint:* Use the data table option for selling price, and determine the selling price at which the NPV is approximately zero.)

C. As the department manager, does this investment seem justified? Explain.

D. From the company's perspective, is this investment justified? What other factors not addressed in the case should be considered in answering this question?

Name:__________________________

CASE 4. MIDWEST TELEPHONE COMPANY SOLUTION

Template: CAPBUDGT

DATA ANALYSIS SUMMARY (attach model printouts as supporting documents):

Midwest Telephone Company

Computer Acquisition – Tax Effects

REQUIREMENT	DEPRECIATION METHOD	NET PRESENT VALUE	IRR	PAYBACK
A1	SL			
A2	DDB5			
A3	DDB7			
A4	DDB5, ITC			
B (Price = $)	DDB5		NR	NR

NR = not required.

DISCUSSION

TUTORIAL 10 USING LOOKUP TABLES

Depreciation on the CAPBUDGT template is calculated with a lookup table. Let's see how a lookup table works.

Step 1 Erase the Screen. If a template is displayed on the screen, remove it with the worksheet erase command. This command erases anything that is on the screen, so be sure to save your work before you execute it.

To erase the worksheet

Press **/WEY** (for *Worksheet*, *Erase*, *Yes*)

Result. The screen shows the blank worksheet.

Step 2 Enter Labels. In this short tutorial, you will enter a year, and the template will look up that year in the lookup table and extract the three different depreciation values for that year. Here, enter the labels for the cells in which you will enter the year and functions that refer to the lookup table.

To enter labels

Goto A1, type `Year` and then press **Enter**
Goto A2, type `Straight line` and then press **Enter**
Goto A3, type `DDB5` and then press **Enter**
Goto A4, type `DDB7` and then press **Enter**

Step 3 Widen Column A. As you can see, the label in cell A2 is wider than the column. You can easily widen a single column at any time. As you widen columns with the arrow keys, the current width is indicated on the status line. Use that indicator to set column A to 15 characters wide.

To widen a single column

Goto any row in column A
Press **/WCS** for (for *Worksheet*, *Column*, *Set*), and the prompt asks you to specify the column width
Press → to widen the column to 15 characters wide. (Press ← to narrow it if you overshoot.)
Press **Enter**

Step 4 Create the Lookup Table. Enter the labels and numbers shown in columns D through G in Figure 33.

Step 5 Name the Lookup Table Range. You will enter three functions in the next step that refer to the lookup table's range. Rather than try to remember the cell coordinates, you can name the table and then use that name in the functions. Here you will also use the arrow keys to point to the range rather than type its cell coordinates.

FIGURE 33 The Lookup Table. After entering the lookup table, your model should look like the one in this figure.

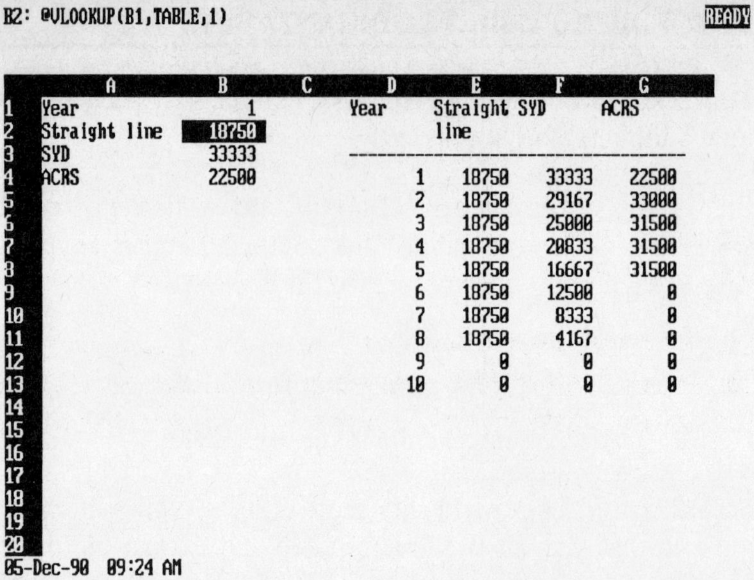

B2: @VLOOKUP(B1,TABLE,1) READY

	A	B	C	D	E	F	G
1	Year	1		Year	Straight	SYD	ACRS
2	Straight line	18750			line		
3	SYD	33333		----------	----------	----------	----------
4	ACRS	22500		1	18750	33333	22500
5				2	18750	29167	33000
6				3	18750	25000	31500
7				4	18750	20833	31500
8				5	18750	16667	31500
9				6	18750	12500	0
10				7	18750	8333	0
11				8	18750	4167	0
12				9	0	0	0
13				10	0	0	0
14							
15							
16							
17							
18							
19							
20							

05-Dec-90 09:24 AM

To name the lookup table

Goto	D4, press **/RNC** (for *Range*, *Name*, *Create*), and the prompt asks you to enter the name for the range
Type	**TABLE** and then press **Enter**. The prompt asks you to enter the range
Press	→ to expand the highlight out to column H (one column past the DDB7 column)
Press	↓ to expand the highlight down to highlight row 13
Press	**Enter**

Result. No change is obvious on the screen, but if you press **Goto**, type **TABLE** and then press **Enter,** the cell pointer will jump to the upper-left corner of the range.

Step 6 Enter Functions That Refer to the Table. The lookup function has four parts. The first is the function name, in this case, @VLOOKUP, which means vertical lookup. (There is also a @HLOOKUP function.) The next three parts are enclosed in parentheses. The first of these is the value to be looked up. This value can be entered in the function directly, or the function can refer to a cell in which it is entered, as we will do here (cell B1). The second of these parts is the range of the table. In this case, you named the table *TABLE,* so you can use that name in the function instead of typing cell coordinates. The last of these parts is the *offset*. This refers to the column on the table that the function is to look up. Here we will use three different offsets since straight line is in column 1, DDB5 in column 2, and DDB7 in column 3.

To check your work, you first enter 1 in cell B1. When you enter functions referring to this cell, the function should display the number on year 1 in the appropriate column.

To enter lookup functions

Goto	B1, type `1` and then press **Enter**
Goto	B2, type `@VLOOKUP(B1,TABLE,1)` and then press **Enter**
Goto	B3, type `@VLOOKUP(B1,TABLE,2)` and then press **Enter**
Goto	B4, type `@VLOOKUP(B1,TABLE,3)` and then press **Enter**

Result. Try changing values in cell B1. Enter any year from 1 to 10, and the three cells below will display the appropriate data from the lookup table.

Step 7 Save Your Table. If you want to save your table, use the program's save commands. Before proceeding, insert your *Cost Accounting Resource Disk* into drive B (on a floppy disks system) or drive A (on a hard disk system).

To save your table

Press	**/FS** (for *File*, *Save*), and the prompt asks you to enter the name of the file to be saved
Type	**`B:TABLE`** (on a floppy disk system) or **`A:TABLE`** on a hard disk system), and then press **Enter**

Result. The drive spins as the file is saved onto the disk.

Step 8 Retrieve AUTO123. If you want to quit, you can press **/QY** (for *Quit*, *Yes*) to return to the operating system. If you want to continue, you can retrieve the AUTO123 file.

To retrieve AUTO123

Press	**/FR** (for *File*, *Retrieve*), and the prompt asks you to enter the name of the file to be retrieved
Type	**`B:AUTO123`** (on a floppy disk system) or **`A:AUTO123`** (on a hard disk system), and then press **Enter**

EXERCISE 11
INVENTORY MANAGEMENT

In this exercise, you solve cases that are based on the principles of economic order quantities discussed in Chapter 24 of *Cost Accounting: A Managerial Emphasis* (8th edition). This exercise assumes that the template EOQ is displayed on the screen. (If you need help on retrieving files, refer to Exercise 1 or Appendix C.)

UNDERSTANDING THE TEMPLATE

The EOQ template has three parts (Figure 34 and Table 12).

Part 1. Calculating the Economic Order Quantity contains two sections. You enter variables into section A, and EOQ-related values are calculated in section B.

Part 2. Graphing the Economic Order Quantity calculates EOQ-related values for a range of order sizes. The data in this part can be displayed on the screen as an EOQ chart if your computer is equipped to display graphics.

Part 3. Sensitivity of EOQ to Variation in Ordering & Carrying Costs calculates EOQ for a range of ordering and carrying costs.

THE TEMPLATE'S GRAPHS

The EOQ template has two graphs.

Graph 1. EOQ plots the carrying, ordering, and total costs against the order sizes specified in Part 2.

Graph 2. SENSITIVITY plots EOQs for ranges of carrying costs and ordering costs specified in Part 3.

FIGURE 34 The EOQ Template.

```
         A           B           C           D           E           F
 1 PART 1: CALCULATING THE ECONOMIC ORDER QUANTITY
 2 =========== =========== =========== =========== =========== ===========
 3
 4 A.  Enter Data
 5 --------------
 6 Enter the number of items used or sold annually (D)           #
 7 Enter annual carrying cost per unit (C)                       $
 8 Enter cost per order (P)                                      $
 9
10 B.  ECONOMIC ORDER QUANTITY CALCULATIONS
11 ---------------------------------------
12 The economic order quantity (EOQ) is:                                    NA
13 The number of orders placed per year is:                                 NA
14
15 Cost of ordering inventory at EOQ:                                       NA
16 Cost of carrying inventory at EOQ:                                       NA
17                                                                 -----------
18      Total                                                               NA
19
20 =========== =========== =========== =========== =========== ===========
21 PART 2: GRAPHING THE ECONOMIC ORDER QUANTITY
22 =========== =========== =========== =========== =========== ===========
23 Enter starting order size                                     #
24 Enter order size increment                                    #
25 =========== =========== =========== =========== =========== ===========
26       Order                             Carrying    Ordering       Total
27        size                                costs       costs       costs
28 ----------- ----------- ----------- ----------- ----------- -----------
29          NA .......... ..........            NA          NA          NA
30          NA .......... ..........            NA          NA          NA
31          NA .......... ..........            NA          NA          NA
32          NA .......... ..........            NA          NA          NA
33          NA .......... ..........            NA          NA          NA
34          NA .......... ..........            NA          NA          NA
35          NA .......... ..........            NA          NA          NA
36          NA .......... ..........            NA          NA          NA
37          NA .......... ..........            NA          NA          NA
38          NA .......... ..........            NA          NA          NA
39          NA .......... ..........            NA          NA          NA
40 =========== =========== =========== =========== =========== ===========
41 PART 3. SENSITIVITY OF EOQ TO VARIATION IN ORDERING & CARRYING COSTS
42 =========== =========== =========== =========== =========== ===========
43                                                       Start   Increment
44 Enter Ordering Costs                          $           $
45 Entering Carrying Costs                       $           $
46 =========== =========== =========== =========== =========== ===========
47    Ordering    [------------------ Carrying Costs ----------------]
48       Costs          NA          NA          NA          NA          NA
49 ----------- ----------- ----------- ----------- ----------- -----------
50          NA          NA          NA          NA          NA          NA
51          NA          NA          NA          NA          NA          NA
52          NA          NA          NA          NA          NA          NA
53          NA          NA          NA          NA          NA          NA
54          NA          NA          NA          NA          NA          NA
55          NA          NA          NA          NA          NA          NA
56          NA          NA          NA          NA          NA          NA
57          NA          NA          NA          NA          NA          NA
58          NA          NA          NA          NA          NA          NA
59          NA          NA          NA          NA          NA          NA
60 =========== =========== =========== =========== =========== ===========
```

Table 12
THE EOQ TEMPLATE

Row or Column	Description
Part 1. Calculating the Economic Order Quantity	
6-8	Enter the input values to calculate EOQ in these rows.
12	The economic order quantity (EOQ) is calculated by the EOQ formula expressed in 1-2-3's syntax — @SQRT(2*F8*F6/F7).
13	The number of orders placed per year is calculated by dividing the number of items sold or used annually (cell F6) by the economic order quantity (cell F12).
15	Cost of ordering inventory at EOQ is calculated by multiplying the cost per order (cell F8) times the number of orders placed per year (cell F13).
16	Cost of carrying inventory at EOQ is calculated by dividing the economic order quantity (cell F12) by 2 and multiplying the result times the annual carrying cost per unit (cell F7).
18	Total is the sum of cost of ordering inventory at EOQ (cell F15) and the cost of carrying inventory at EOQ (cell F16).
Part 2. Graphing the Economic Order Quantity	
A	Order size is set by the starting order size (cell F23) and the order size increment (cell F24).
D	Carrying costs are calculated by multiplying the annual carrying cost per unit (cell F7) times one-half the order size in column A on the same row.
E	Ordering costs are calculated by multiplying the cost per order (cell F8) times the result of dividing the number of items sold or used (cell F6) by the order size in column A on the same row.
F	Total costs are calculated by adding carrying costs (column D) and ordering costs (column E) on the same row.
Part 3. Sensitivity of EOQ to Variation in Ordering & Carrying Costs	
48	Carrying costs are specified by the start (cell E45) and increment (cell F45) values.
A	Ordering costs are specified by the start (cell E44) and increment (cell F44) values.
B-F	EOQ is calculated for each pair of ordering and carrying cost values.

CASE 1 BILL'S DISCOUNT STORES

Bill's Discount Stores (BDS) is deciding on the purchase order quantity for B&D electric drills. A review of past sales indicates that annual demand for these electric drills is 10,000 units.

A recent study by Sara Yost, Assistant Controller for BDS, indicated that the cost of processing a purchase order was $25.00. As part of this study, Sara also measured the inventory carrying cost for small tools, such as this electric drill, and concluded that it averaged $2.00 per unit per year, including the opportunity costs.

BDS uses the standard economic-order-quantity (EOQ) model in deciding on its purchase order size.

REQUIREMENTS

A. Calculate the total cost of purchase order quantities, beginning with 250 and incrementing by 50 (use Part 2 of the template), for the electric drill that BDS plans to purchase. Identify the EOQ, and comment on your results.

B. Assume that the carrying cost was misestimated and should have been $3.00 per unit per year. What is the cost of this estimation error? (*Hint:* Recalculate the template with the new carrying cost estimate and compare the results with Requirement A.)

C. Return to the $2.00 per unit per year carrying cost estimate. Substitute *decreasing* ordering costs into the template until the EOQ declines to *100* and print the final template.

D. Return to the $25.00 cost per order. Substitute *increasing* carrying costs into the template until the EOQ declines to *100* and print the final template.

E. Explain to management, using the results of Requirements C and D, why many companies are adopting Just-In-Time (JIT) purchasing.

Name: ______________________

CASE 1. BILL'S DISCOUNT STORES
Template: EOQ

DATA ANALYSIS SUMMARY (attach model printouts as supporting documents):

Bill's Discount Stores
EOQ Analysis – B&D Electric Drills

REQUIREMENT	COST PER ORDER	UNIT CARRYING COSTS	EOQ
A	$25	$2	
B	25	3	
C		2	100
D	25		100

DISCUSSION

CASE 2 COLUMBUS STORES

The management of Columbus Stores has heard of the advantages of JIT purchasing. They request a recommendation on the adoption of this purchasing procedure from the accounting staff.

You decide to select a typical product as the basis for the analysis to support your recommendation. The product, automatic coffee makers, has an annual demand of 10,000 units. It costs $50.00 to process a purchase order and $4.50 to carry one unit for one year.

REQUIREMENTS

A. Calculate the economic order quantity (EOQ) and total cost of carrying and ordering the coffee makers using the current purchasing system.

B. Using Part 3 of the template, calculate the EOQs for the coffee makers under the following conditions:

1. Ordering costs of $50.00 reduced in $5.00 increments, and
2. Carrying costs of $4.50 per unit per year increased in $1.00 increments.

C. Prepare a recommendation to management. Include in your recommendation some suggestions for reducing ordering costs and some reasons that carrying costs are typically underestimated.

Name:______________________________

CASE 2. COLUMBUS STORES
Template: EOQ

DATA ANALYSIS SUMMARY (attach model printouts as supporting documents):

Columbus Stores
EOQ Analysis – Automatic Coffee Makers

REQUIREMENT	EOQ	TOTAL COST
A		

DISCUSSION

CASE 3 VALLEY MANUFACTURING COMPANY

Valley Manufacturing Company (VMC) produces several models of store fixtures. One advantage they have over their competition is that they can provide faster delivery because they keep a substantial inventory of every model in their catalog. The management of VMC does not want to lose this competitive edge but wonders whether there is some way of reducing this costly finished goods inventory by tying production closer to customer orders.

Sam Sloth, the manager of production planning for the last 15 years, argues that this cannot be done. He explains that tying production to sales would require frequent production changeovers from model to model. This would be very costly because of the equipment setup costs necessary to change over from producing one model to another. Sam favors the policy VMC has been following, which is to have long production runs of each model. This way, he argues, you can average the substantial setup cost over many units of the particular model being run.

Don Aggressive, the corporate controller, decides to become part of this debate once he hears Sam's claim about averaging the substantial setup costs. First, Don selects Model 75, a high-volume, standard-sized store fixture for analysis. A review of the records indicates that annual demand for Model 75 is 50,000 units. The carrying cost of Model 75 is $25 per unit per year. Don initiates a cost study of the setup costs to manufacture Model 75 and concludes that it costs $900 per setup.

Don is also aware of the work of Larson Consulting, which has studied setup costs extensively and achieved substantial savings for their clients. Articles Don has read report that setup costs have routinely been reduced by 80%.

REQUIREMENTS

A. Calculate the current economic lot size (ELS) for Model 75 and the total cost of this policy. The ELS formula is the same as the EOQ formula except that setup costs are substituted for ordering costs.

B. Calculate the total cost of production quantities beginning with 500 and incrementing by 500 (use Part 2 of the template) for the current $900 setup costs.

C. Calculate the ELS if setup costs could be reduced to $100 and the total cost of this policy.

D. Calculate the total cost of production quantities beginning with 200 and incrementing by 100 (use Part 2 of the template) for the $100 setup costs.

E. Using the results of the previous requirements, present a proposal to management recommending shorter production runs.

Name: ______________________

CASE 3. VALLEY MANUFACTURING COMPANY

Template: EOQ

DATA ANALYSIS SUMMARY (attach model printouts as supporting documents):

Valley Manufacturing Company
EOQ Analysis – Model 75

REQUIREMENT	ELS	TOTAL COST
A		
D		

DISCUSSION

TUTORIAL 11 EXPLORING FORMATS

In this tutorial, you use the global and range format commands. Your finished model will look like Figure 35. Refer to this figure to check your results as you complete the steps in this tutorial.

Step 1 Erase the Screen. If a template is displayed on the screen, remove it with the worksheet erase command. This command erases anything that is on the screen, so be sure to save your work before you execute it.

To erase the worksheet

Press **/WEY** (for *Worksheet*, *Erase*, *Yes*)

Result. The screen shows the blank worksheet.

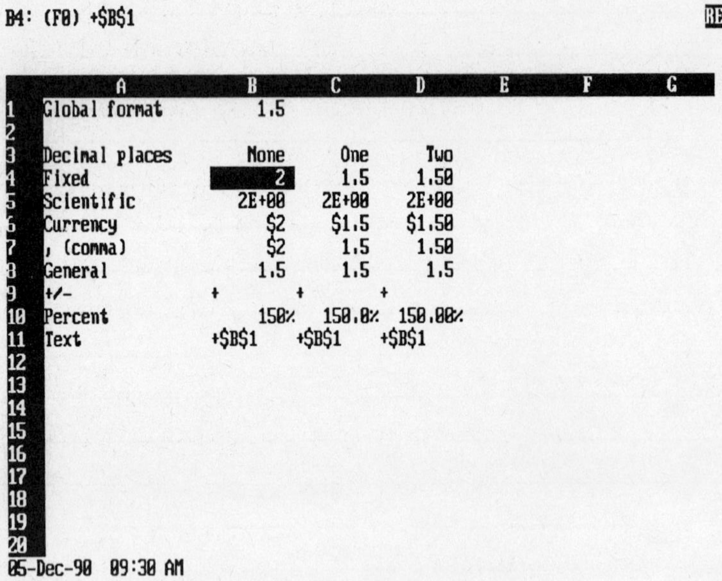
B4: (F0) +$B$1 READY

	A	B	C	D	E	F	G
1	Global format	1.5					
2							
3	Decimal places	None	One	Two			
4	Fixed	2	1.5	1.50			
5	Scientific	2E+00	2E+00	2E+00			
6	Currency	$2	$1.5	$1.50			
7	, (comma)	$2	1.5	1.50			
8	General	1.5	1.5	1.5			
9	+/-	+	+	+			
10	Percent	150%	150.0%	150.00%			
11	Text	+$B$1	+$B$1	+$B$1			
12							
13							
14							
15							
16							
17							
18							
19							
20							

05-Dec-90 09:30 AM

FIGURE 35 The FORMAT Model. After completing this tutorial, your model should look like the one in this figure.

Step 2 Enter Labels. Begin by entering the labels shown in column A and on row 3 of Figure 35. Don't forget to enter a label-prefix character for the +/- label. Also use the label-prefix " for the labels *None*, *One*, and *Two* so that they are right-aligned in their cells.

Step 3 Change Column Widths. If you look closely at the labels you entered in column A, you notice that many of them are wider than the column. You can easily widen the column so that the labels are not truncated when you enter data in the next column.

To widen column A

Press **Home** to move the cell pointer to column A

Press **/WCS** (for *Worksheet*, *Column*, *Set-Width*), and the prompt asks you to specify a column width

Press → to scroll the column wider, or press ← to scroll it narrower

Result. As you press → or ←, the number following the prompt indicates the current column width. Scroll the column so that the prompt indicates it is fifteen characters wide, and then press **Enter**.

Step 4 Enter a Number and Formulas. To see the effect of different formats on the same number, you will enter a number in cell B1. You will then enter a formula that refers to this number, and copy it to other cells on the model. This way, when you change the number, all other numbers change automatically. The cell reference that you enter contains dollar signs. These are used to specify an absolute reference that determines the cell the formula refers to when copied.

To enter a number

Goto	B1 type `1.5` and then press **Enter**

To enter a formula

Goto	B4 type `$B$1` and then press **Enter**

To copy the formula

Goto	B4 press **/C** (for *Copy*), and the prompt asks you to *enter a range to copy FROM:*
Press	**Enter** to copy from the cell containing the cell pointer, and the prompt asks you to *enter a range to copy TO*:
Type	`B4.D11` and then press **Enter**

Result. All the cells in the model display the same value.

Step 5 Change the Global Format. Before changing range formats, change the global format a few times to see how it affects all cells. As you make the changes, notice the results each format has.

To change the global format

Press	**/WGFC** (for *Worksheet*, *Global*, *Format*, *Currency*), and the prompt asks for the number of decimal places
Press	**Enter** to accept the suggested setting
Press	**/WGFP** (for *Worksheet*, *Global*, *Format*, *Percent*), and the prompt asks for the number of decimal places
Type	`0` and then press **Enter**
Press	**/WGFG** (for *Worksheet*, *Global*, *Format*, *General*)

Result. Each time you change the global format, all numbers on the model change formats.

Step 6 Format Individual Cells. Now, use a range format to format each number in the model (except the global number in cell B1). We explain only the first format. To do the others, move the cell pointer to each cell, and use the format specified in column A and the number of decimal places on row 3.

Note. The General, +/-, and Text formats, unlike the others, do not prompt you for the number of decimal places.

To format a cell

Goto	B4 press **/RFF** (for *Range*, *Format*, *Fixed*), and the prompt asks for the number of decimal places
Type	`0` and then press **Enter**. The prompt asks for the range to format
Press	**Enter** to format only the cell containing the cell pointer

Result. After you have formatted each of the numbers, your model should match the one in Figure 35.

Step 7 Change the Number. Now you have a model you can use to experiment with. Enter numbers like 10, 100, 1000, 10000, and 100000 in cell B1, and see the effects the size of the number has on the numbers displayed on the model. When one or more cells displays a row of asterisks, try widening the columns as described in Step 2, or change the global column width by pressing **/WGC** (for *Worksheet*, *Global*, *Column-Width*). Press → and ← to adjust the column widths, and then press **Enter**. Do not bother to widen the columns to remove the asterisks from the +/- row since the number of characters is directly related to the number you enter in cell B1. For example, if you enter the number 100, you have to set the column width to 100 to display the results on this row.

Step 8 Change the Global Format. Change the global format a few times. Notice how the cells you formatted with a range format do not respond? Only the number in cell B1, which you did not format with a range format, changes. This is because range formats have precedence over global formats.

Step 9 Print the Model. Be sure that the printer is on and has paper in it and that the paper is aligned. Then make a printout of the model, advance the paper out of the printer, and leave the Print menu.

Step 10 Save the Model. Now that you have finished the model, save it under the filename FORMAT so that you can retrieve it later. Before proceeding, insert your *Cost Accounting Resource Disk* into drive B (on a floppy disks system) or drive A (on a hard disk system).

To save the model

Press	**/FS** (for *File*, *Save*) and the prompt asks you to enter the name of the file
Type	**`B:FORMAT`** on a hard disk system or **`A:FORMAT`** (on a hard disk system), and then press **Enter**

Result. The file is now saved onto the disk.

Step 11 Retrieve AUTO123. If you want to quit you can press **/QY** (for *Quit*, *Yes*) to return to the operating system. If you want to continue you can retrieve the AUTO123 file.

To retrieve AUTO123

Press	**/FR** (for *File*, *Retrieve*) and the prompt asks you to enter the name of the file to be retrieved
Type	**`B:AUTO123`** (on a floppy disk system) or **`A:AUTO123`** (on a hard disk system), and then press **Enter**

EXERCISE 12
MANAGING DIVISION ROI

In this exercise, you solve cases that are based on the principles of return on investment discussed in Chapter 26 of *Cost Accounting: A Managerial Emphasis* (8th edition). This exercise assumes that the template ROI is displayed on the screen. (If you need help on retrieving files, refer to Exercise 1 or Appendix C.)

UNDERSTANDING THE TEMPLATE

The ROI template has four parts (Figure 36 and Table 13). You enter data into Parts 1 and 2, and Part 3 calculates ROI and its components using current and historical cost measures. Part 4 can be used to explore the effects of changes in sales, costs, and assets on ROI.

Part 1. Invested Capital is where you enter historical cost balance sheet data and indexes of construction costs.

Part 2. Operating Income is where you enter historical cost income statement data.

Part 3. ROI calculates the operating margin, asset turnover, and ROI based on the data you enter in Part 1.

Part 4. Effects of Changes in Single Variables allows you to change each individual variable — sales, costs, and assets — by a percentage you define and observe the impact of the changes on ROI.

Name:______________________________

CASE 2. FARNAM DEPARTMENT STORES SOLUTION
Template: ROI

DATA ANALYSIS SUMMARY (attach model printouts as supporting documents):

Farnam Department Stores
Store Performance
For the Year Ended February 29, 19x1

REQUIREMENT (STORE)	METHOD	OPERATING MARGIN	ASSET TURNOVER	ROI
A (Springfield)	Hist. Cost-Net			
A (Lincoln)	Hist. Cost-Net			
A (Cedar Falls)	Hist. Cost-Net			
B (Springfield)	Curr. Cost-Net			
B (Lincoln)	Curr. Cost-Net			
B (Cedar Falls)	Curr. Cost-Net			

DISCUSSION

FIGURE 36 The ROI Template.

	A	B	C	D	E	F	G
1	PART 1. INVESTED CAPITAL						
2	============	============	============	============	============	============	
3				Historical		Current	
4				cost		cost	
5	------------	------------	------------	------------	------------	------------	
6	Current assets			$		NA	
7				------------		------------	
8							
9	Gross fixed assets			$		NA	
10	Less accumulated depreciation			$		NA	
11				------------		------------	
12	Net fixed assets			0		NA	
13				------------		------------	
14	Total invested assets			0		NA	
15				============		============	
16	Index of construction costs:						
17	------------	------------	---				
18	Beginning year		#				
19	Current year		#				
20	============	============	============	============	============	============	
21	PART 2. OPERATING INCOME						
22	============	============	============	============	============	============	
23			[-- Historical cost --]		[--- Current cost ---]		
24			$	%	$	%	
25	------------	------------	------------	------------	------------	------------	
26	Sales		$	NA	$0	NA	
27	Variable costs		$	NA	0	NA	
28			------------	------------	------------	------------	
29	Contribution margin		0	NA	0	NA	
30			------------	------------	------------	------------	
31	Fixed Costs:						
32	------------						
33	Operating		$	NA	0	NA	
34	Depreciation		$	NA	0	NA	
35			------------	------------	------------	------------	
36	Total fixed costs		0	NA	0	NA	
37			------------	------------	------------	------------	
38	Operating income		0	NA	0	NA	
39							
40	============	============	============	============	============	============	

FIGURE 36 (Cont.)

```
          A           B           C           D           E           F           G
41 PART 3. ROI
42 ============ ============ ============ ============ ============ ============
43 [FORMULA]
44              Operating margin   X  Asset turnover     =       ROI
45
46               Net income                   Sales                  Net income
47              ------------       X      ------------     =      ------------
48                 Sales                   Invested                 Invested
49                                          capital                  capital
50
51 ============ ============ ============ ============ ============ ============
52 [CALCULATIONS]
53                                  Operating                   Asset
54 Method                              margin                 turnover         ROI
55 ------------ ------------ ------------ ------------ ------------ ------------
56 Historical cost - Gross                 NA                       NA          NA
57 Historical cost - Net                   NA                       NA          NA
58 Current cost - Gross                    NA                       NA          NA
59 Current cost - Net                      NA                       NA          NA
60 ============ ============ ============ ============ ============ ============
61 PART 4. EFFECTS OF CHANGES IN SINGLE VARIABLES
62 ============ ============ ============ ============ ============ ============
63 Enter 1 for Historical cost-Net; or 2 for Current cost-Net.:   #
64 Enter increment (- for decreases)............:   Sales                   %
65                                                  Costs                   %
66                                                  Assets                  %
67 ============ ============ ============ ============ ============ ============ ============
68        Sales          ROI        Costs          ROI       Assets          ROI     Original
69 ------------ ------------ ------------ ------------ ------------ ------------ ------------
70           NA           NA           NA           NA           NA           NA           NA
71           NA           NA           NA           NA           NA           NA           NA
72           NA           NA           NA           NA           NA           NA           NA
73           NA           NA           NA           NA           NA           NA           NA
74           NA           NA           NA           NA           NA           NA           NA
75           NA           NA           NA           NA           NA           NA           NA
76           NA           NA           NA           NA           NA           NA           NA
77           NA           NA           NA           NA           NA           NA           NA
78           NA           NA           NA           NA           NA           NA           NA
79           NA           NA           NA           NA           NA           NA           NA
80 ============ ============ ============ ============ ============ ============ ============
```

Table 13
THE ROI TEMPLATE

Row or Column	Description
Part 1. Invested Capital	
D	Historical cost values for current assets, gross fixed assets, and accumulated depreciation are entered in this column.
F	Current costs are calculated in this column. Gross fixed assets (cell F9) are adjusted based on the index of construction costs you enter on rows 18 and 19. Accumulated depreciation (cell F10) is historical cost accumulated depreciation (cell D10) divided by historical cost gross fixed assets (cell D9). The resulting percentage is multiplied times current gross assets (cell F9).
18 & 19	You enter the beginning and current year's index of construction costs on these rows.
Part 2. Operating Income	
C	Historical cost sales, variable costs, and fixed costs are entered in the appropriate rows in this column.
D	The values in column C are reduced to a percentage of sales in this column.
E	Current cost values are calculated in this column. Sales, variable costs, and operating fixed costs are the same as those in column C. Depreciation is adjusted based on the index of construction costs you entered on rows 18 and 19.
F	The values in column E are reduced to a percentage of sales in this column.
Part 3. ROI	
43-50	The ROI formula is shown to guide you.
C	Historical cost operating margins (rows 56 and 57) are historical cost operating income (cell C38) divided by historical cost sales (cell C26). Current cost operating margins (rows 58 and 59) are current cost operating income (cell E38) divided by current cost sales (cell E26).
E	Asset turnover is sales (cell C26) divided by the appropriate asset base: historical cost net is historical cost invested assets (cell D14); historical cost gross is historical cost net plus accumulated depreciation (cell D10); current cost net is current cost invested assets (cell F14); current cost gross is current cost net plus accumulated depreciation (cell F10).
F	ROI is operating margin (column C) times asset turnover (column E).
Part 4. Effects of Changes in Single Variables	
63	You specify historical cost-net or current cost-net on this row.
64-66	You enter percentages by which sales, costs, and assets are to be changed. To see the effects of increases, enter positive percentages (for example, .1); to see the effects of decreases, enter negative percentages (for example, -.1).
A	The range of sales start at the value on row 26 and increment by the percentage you enter on row 64.
B	ROI is calculated for each level of sales in column A based on either historical cost-net or current cost-net as specified on row 63. Variable costs are assumed to change proportionately.
C	The range of costs starts at one of two values. If you specified historical cost-net on row 63, it starts with the sum of historical variable costs (cell C27) plus historical total fixed costs (cell C36). If you specified current cost-net, it uses the current costs in column E from the same rows. The subsequent rows increment by the percentage you enter on row 65.
D	ROI is calculated for each level of costs in column C and is based on either historical cost-net or current cost-net as specified on row 63.

Table 13 (Cont.)

E	Assets start at the values entered on row 14. They are carried from the historical or current cost columns on that row depending on your entry on row 63. The subsequent rows increment by the percentage you enter on row 66.
F	ROI is calculated for each level of assets in column E and is based on either historical cost-net or current cost-net as specified on row 63.

THE TEMPLATE'S GRAPHS

The ROI template has three graphs that plot the changes in single variables from Part 4. Each graph also plots the original ROI before changes were made in single variables for comparison purposes.

Graph 1. ROI: Changes in Sales illustrates how changes in sales affect ROI. It plots the revised ROI (column B) against sales (column A) and the original ROI before changes were made in sales.

Graph 2. ROI: Changes in Costs illustrates how changes in costs affect ROI. It plots the revised ROI (column D) against costs (column C) and the original ROI before changes were made in costs.

Graph 3. ROI: Changes in Assets illustrates how changes in assets affect ROI. It plots the revised ROI (column F) against assets (column E) and the original ROI before changes were made in assets.

CASE 3 DODGE ENTERPRISES

The Custom Software Division of Dodge Enterprises is not performing up to expectations. Even though it is a relatively new division, considerable resources were committed to it because the market research indicated substantial growth potential. By now, its third year of operation, management had expected a return on investment (ROI) of 25% calculated on net book value measured with historical cost. The division's operating performance for last year is reported below:

Custom Software Division
Operating Income
For the Year Ended June 30, 19x1

Revenues		$450,000
Variable Expenses		200,000
Contribution Margin		$250,000
Fixed Expenses		
Operating (excluding depreciation)	$180,000	
Depreciation	30,000	210,000
Operating Income		$ 40,000

The current assets controlled by the division total $100,000 and are primarily accounts receivable. Much of the division's equipment is leased, so the gross book value of fixed assets totals only $120,000 with accumulated depreciation to date of $24,000.

A recent market survey indicates that the market potential for the division is still strong. Although there are more competitors than there were three years ago, the competition is not any greater than was predicted when the division was formed.

REQUIREMENTS

A. Compute the ROI for the Custom Software Division. By how much did they miss the ROI target?

B. Holding everything else constant, calculate the ROIs for 1% increments in revenues. Use Part 4 of the template. Approximately what amount of revenues is necessary to meet the ROI target? Describe how this change affects ROI.

C. Holding everything else constant, calculate the ROIs for .5% decreases in costs. (Remember to enter .5% as -.005.) Use Part 4 of the template. Approximately what would costs have to be to meet the ROI target? Describe how this change affects ROI.

D. Holding everything else constant, calculate the ROIs for 4% decreases in assets. (Remember to enter 4% as -.04.) Use Part 4 of the template. What would assets have to be to meet the ROI target? Describe how this change affects ROI.

E. What operating changes would you recommend to assist the division in meeting its ROI target?

Name: ______________________

CASE 3. DODGE ENTERPRISES SOLUTION

Template: ROI

DATA ANALYSIS SUMMARY (attach model printouts as supporting documents):

Custom Software Division
Performance Review and Analysis
For the Year Ended June 30, 19x1

REQUIREMENT	METHOD	REVENUE	COSTS	ASSETS	ROI
A	Hist.Cost-Net	$450,000	$410,000	$196,000	
B	Hist.Cost-Net			196,000	
C	Hist.Cost-Net	450,000		196,000	
D	Hist.Cost-Net	450,000	410,000		

DISCUSSION

TUTORIAL 12 REVISING GRAPHS AND MENUS

The ROI template has three graphs that show the effect of changes in sales, costs, and assets on ROI. Let's also display one of these bar graphs as a line graph, name it, and then add it to the template menu.

Step 1 Retrieve the ROI Template. Retrieve the template if it isn't already on the screen. Enter the same data you entered for one of the cases. Before doing this, press **/WGPD** (for *Worksheet*, *Global*, *Protection*, *Disable*) to turn off protection.

Step 2 Create a New Graph. Let's create a new graph. To do so, you first enter a new X-axis range since the previous graphs are all plotted against the changing variable. You then reset the current graph settings so that they don't interfere with the new graph.

To enter a new X-axis

Goto	H70, type `1` and then press **Enter**
Goto	H71, type `1+H70` and then press **Enter**
Press	**/C** (for *Copy*), and the prompt asks you to enter the range to copy FROM
Press	**Enter**, and the prompt asks you to enter the range to copy TO
Type	`H72.H79` and then press **Enter**

To reset the graph settings

Press	**/GRG** (for *Graph*, *Reset*, *Graph*)

To set the X-axis for the new graph

Press	**X** (for *X*), and the prompt asks you to enter the X-axis range
Type	**`H70.H79`** and then press **Enter**

To enter the data ranges

Press	**A** and the prompt asks you to enter the first data range
Type	**`B70.B79`** and then press **Enter**
Press	**B** and the prompt asks you to enter the second data range
Type	**`D70.D79`** and then press **Enter**
Press	**C** and the prompt asks you to enter the third data range
Type	**`F70.F79`** and then press **Enter**

To select the graph type

Press	**TL** (for *Type*, *Line*)
Press	**V** (for *View*)

Result. The new graph appears on the screen. When finished looking at the graph, press **Spacebar** to clear the graph and return to the Graph menu.

Step 3 Name the Graph. You can save a "library" of graphs by assigning them names.

To name the graph

Press	**NC** (for *Name*, *Create*), and the prompt asks you to enter the name of the graph
Type	`GRAPH 4` and then press **Enter**
Press	**Q** (for *Quit*) to remove the Graph menu

Result. Creating and naming the graph added it to the library of graphs stored by the program.

Step 4 Add the Graph to the Menu. Now that the graph has been added to the library of named graphs, you can add it to the menu.

To add the graph to the menu

Press	**Goto** (**F5** on the IBM PC), and the prompt asks you to enter the address of the cell to go to
Type	`N21` and then press **Enter**

Result. The macro for the assets graph is displayed on the screen.

To copy the assets graph macro

Press	**/C** (for *Copy*), and the prompt asks you to enter the range to copy FROM
Type	`N26.N29` and then press **Enter**. The prompt asks you to enter the range to copy TO
Type	`O26` and then press **Enter**

To revise the copy of the graph 3 macro into a graph 4 macro

Press	**Tab** to move one column to the right
Goto	O26, type `Combined` and then press **Enter**
Goto	O27, type `Display combined graph` and then press **Enter**
Goto	O28, type `/gnuGRAPH 4~q` and then press **Enter**
Type	`xmMENU2~`

FIGURE 37 Macro That Displays a Graph. After entering your macro, your screen should look like the one in this figure.

Result. Your macro should look like Figure 37. When you have finished, you have added the graph to the menu. To see it, press **Alt-M** to display the template menu, and then press **G** (for *Graphs*). *Combined* appears on the menu. Move the menu pointer over it, and you see the second line *Display combined graph*. Press **Enter** to display the new graph. When you have finished, press **Spacebar** to return to the template. To save the revised template, press **B** (for *Backup*), and then press **S** (for *Save*). This is the last tutorial, you can quit the program or retrieve AUTO123 if you want to retrieve another template.

APPENDIXES

A. Backing Up Your Original Cost Accounting Resource Disk

You should make a backup copy of the original *Cost Accounting Resource Disk* in case you inadvertently erase or alter one of the template files. If you do, you can copy the file from the original disk onto the backup copy and continue working.

To complete the steps in this section, you need the following:

1. A copy of the MS-DOS or PC-DOS disk.
2. A copy of the Cost Accounting Resource Disk.
3. One or two blank disks.
4. Two blank labels for your disks.
5. Two pieces of tape to cover the write-protect notches on the DOS disk and the original *Cost Accounting Resource Disk* if you are using 5¼-inch disks.

Step 1 Label and Write-Protect Your Disks. There is an unwritten rule among computer users that an unlabeled disk does not contain valuable files. No one has the time to check an unlabeled disk to see what files, if any, it contains. For this reason, the first step is always to label your disk. Write your name, date, format (for example, DOS 2.0), and *Cost Accounting Resource Disk-Backup* on one disk label and *Data Disk* on the other disk label if you are formatting an extra disk.

Write-protecting a disk ensures that its files are not erased or damaged by mistake. If you are using 5¼-inch disks, cover the write-protect notches with write-protect tape on the DOS and the original *Cost Accounting Resource Disk*. If you are using 3½-inch disks, open their write-protect windows.

Step 2 Load DOS. To use MS-DOS or PC-DOS, the disk operating system must be loaded into the computer's memory.

1. Hold the disk, labeled DOS, with the label facing up. (If the disk is in your right hand and your thumb is on the label, you are holding it correctly.)
2. Open the door to disk drive A. (It is the one on the left if the drives are side by side or the one on top if they are one above the other.)
3. Hold the disk with the oblong opening pointing toward the slot in the drive. Insert the disk gently into the drive. Never push hard because the disk will buckle if it gets caught on an obstruction. Carefully jiggle the disk to make sure it is inserted all the way into the disk drive.
4. Gently close the disk drive door. If it won't close all the way, open the door, jiggle the disk, and close the door again.
5. Boot the system. If the computer is off, turn it on. If the computer is already on, hold down **Ctrl** and **Alt** while you press **Del**. (Anything displayed on the screen will be erased by this command. If another program is displayed, check that the previous user didn't just go to get a cup of coffee or you may lose a friend when he or she returns to find you have erased an important file.)

6. If you have just turned on the computer and the computer does not have a clock that is set automatically, you will be asked to enter the date and time. Enter the date as month-day-year, separated by hyphens (for example, 01-30-92 for January 30,1992), and then press **Enter**. Enter the time as hour:minute separated by a colon (for example, 1:30), and then press **Enter**.
7. The operating system appears on the screen.

Step 3 Format a Data Disk. You must format a new disk before you save files onto it. Follow these steps to format a new disk:

1. Insert the blank disk labeled *Cost Accounting Resource Disk-Backup* in drive B (on a floppy disk system) or drive A (on a hard disk system).
2. Type `FORMAT B:` (on a floppy disk system) or `FORMAT A:` (on a hard disk system), and then press **Enter**. The drive spins for a moment. (Never open the drive doors while the drive is spinning and the drive's light is on. This can cause damage to the files on the disk.) In a moment, a prompt asks you to *Insert new diskette for drive x: and strike any key when ready.*
3. Press any key. The drive spins as it formats the disk. In a minute or so, the drive stops, and the prompt tells you that *Formatting...Format complete* and then asks *Format another (Y/N)?.*
4. If you want to format another disk, press **Y**. To stop, press **N** and the *A>* prompt reappears.

Step 4 Copy Files. The way you copy files depends on the system you are using:

- On a floppy disk system, insert the original *Cost Accounting Resource Disk* into drive A. Leave the disk you just formatted in drive B.
- On a hard disk system with a single floppy disk drive, insert the original *Cost Accounting Resource Disk* into drive A.

To copy files

Type `COPY A:*.* B:` and then press **Enter**

Result. As files are copied, they are listed on the screen. If you are using a system with a single floppy disk drive, you are periodically prompted to swap disks. When all the files have been copied, the command prompt reappears. Remove the original *Cost Accounting Resource Disk,* and then store it in a safe place. Insert the backup copy into drive B (on a floppy disk system) or drive A (on a hard disk system), and then proceed to Exercise 1 for instructions on loading your program.

B. Using the AUTO123 Template Menu

When you load your program with the *Cost Accounting Resource Disk* in one of the computer's disk drives, the AUTO123 file appears on the screen with its menu displayed automatically. If this does not happen, you can retrieve the file and display its menu manually, or you can change settings so that the file is retrieved and its menu is displayed automatically the next time you load 1-2-3.

Retrieving AUTO123 Manually

1. Insert the backup *Cost Accounting Resource Disk* into drive B (on a floppy disk system) or drive A (on a hard disk system)
2. Press **/FR** (for *File*, *Retrieve*), and the prompt reads *Enter name of file to retrieve:*
3. Type **B:AUTO123** (on a floppy disk system) or **A:AUTO123** (on a hard disk system), and then press **Enter**

Retrieving AUTO123 Automatically

1. Press **/WGDD** (for *Worksheet*, *Global*, *Default*, *Directory*) and the prompt reads *Enter default directory:* followed by the current default setting
2. Press **Esc** to clear the current setting
3. Type **B** (on a floppy disk system) or **A** (on a hard disk system), and then press **Enter**
4. Press **U** (for *Update*)
5. Press **Q** (for *Quit*)

Displaying Template Menus Manually

1. Retrieve the template
2. Hold down **Alt** and press the letter **M**

Displaying Template Menus Automatically

1. Press **/WGDA** (for Worksheet, Global, Default, Autoexec)
2. Press **Y** (for Yes)
3. Press **U** (for Update)
4. Press **Q** (for Quit)

Selecting Choices Listed on the Template Menu

The AUTO123 file is the command center for retrieving templates on the disk accompanying this book. At the top of the screen is a list of choices, called a menu. These choices are described in Table 14. You can select any of these choices in one of two ways:

- Press the first letter in the choice's name, for example, press **R** (for *Retrieve Templates*).
- Press → or ← to move the menu pointer over the desired menu choice, and then press **Enter**.

Table 14
AUTO123 TEMPLATE MENU CHOICES

Menu Choice	Description
Retrieve Templates	Displays a menu listing the first six templates on the disk. • To retrieve one of the templates, press its number. • To display additional template names, press **M** (for *More*). • To return to the previous menu, press **B** (for *Backup*).
Copyright	Displays the copyright notice for the *Cost Accounting Resource Disk*.

Table 14 (Cont.)

Description	Displays a brief description of the *Cost Accounting Resource Disk.*
Quit	Returns you to the operating system so that you can quit for the day or run another program. When you select this choice, follow the prompts that appear on the screen.
Enter Name	Enters your name the first time you use the *Cost Accounting Resource Disk* so that it appears on your printouts (see Step 4 in Exercise 1).
Printer	Requests information needed to print in compressed type (see Step 5 in Exercise 1).

C. Retrieving a Template

Here is a summary of the steps to retrieve one of the templates from the disk. Complete instructions are given in Steps 1 through 4 of Exercise 1, "Getting Acquainted with 1-2-3."

1. Load your Lotus 1-2-3 program with the program disk in drive A and the backup *Cost Accounting Resource Disk* in drive B as described in Appendix B. The AUTO123 file automatically appears on the screen. (If it doesn't, press **/FR** (for *File*, *Retrieve*), type **AUTO123** and then press **Enter.**) The AUTO123 template at the top of the screen lists choices. (If it doesn't, hold down **Alt** and press **M.**) Press **R** (for *Retrieve Templates*) to display a list of template names.
   - To retrieve one of the listed templates, type its number.
   - To display additional template names, press **M** (for *More*).
   - To return to the previous menu, press **B** (for *Backup*).
2. When a template is retrieved, a *WAIT* message appears on your screen as your name is copied to the template. When the drive stops spinning, and the message *Press Enter to continue* is highlighted, press **Enter** to display the template.

When you retrieve the templates for Exercises 5 (CASHFLOW), 9 (PRODCOST), and 10 (CAPBUDGT) and then press **Enter** after your name is copied into the template, the printer setup codes screen appears. The program copies the printer setup codes for compressed and normal type that you entered in Exercise 1. If you didn't enter codes for your printer in Exercise 1, the codes for the Epson printers are copied in automatically. You can change these codes before you print, as described in Appendix F, "Using The Template Menus." When the drive stops spinning, and the prompt *Press Enter to continue* is highlighted, press **Enter** to display the template.

D. Getting Around a Template

All the templates on the *Cost Accounting Resource Disk* (except AUTOLOAN in Exercise 1) have two or more parts, only one of which can be displayed on the screen at a time. The templates, however, are designed so that you can scroll the screen quickly from one part to another.

- To scroll the screen down and up a part at a time, press **PgDn** and **PgUp**.
- To scroll the screen sideways, press **Tab** to move right and hold down **Shift** and press **Tab** to move left. (To keep the row labels from scrolling off the screen when you do this, see Appendix I, "Using Fixed Titles.")
- To go directly to a specific part, press **Goto** (**F5** on the IBM PC), and the prompt reads *Enter address to go to:*. Type the name of the part that you want to see, and then press **Enter**. For example, to move to Part 2, press **Goto**, type **PART 2** and then press **Enter**. If the screen doesn't align properly, press **Home** before pressing **Goto**.
- To realign the screen if you get it out of alignment, press **Home**, and then press **PgDn** to move to the part you want to see.
- 1-2-3 has many keys and commands to move the cell pointer around the worksheet. Table 15 summarizes those that are used most frequently.

Table 15
CELL POINTER MOVEMENT KEYS

Key	Effect On Cell Pointer
→ or ←	Right or left one column
↓ or ↑	Down or up one row
Tab	Right one screen page
Shift-Tab	Left one screen page
PgDn or **PgUp**	Down or up one screen page
Home	To cell A1
Goto [F5]	Used to enter a specific cell address
Scroll-Lock	The cell pointer remains fixed and the screen scrolls

TIP ON THE TAB KEY

On the IBM PC, **Tab** is the key with the symbol →| on it. The same key also has the symbol |← indicating the screen scrolls to the left when you press the key while holding down **Shift**.

E. Entering Data into a Template

If you have never entered data into a template, you will find it quite easy. Just follow these steps:

1. Check that the mode indicator in the upper-right or lower-left corner reads *READY*. If it doesn't, select *Quit-menu* if the indicator reads *CMD* or press **Esc** a few times.
2. Use the keys described in Appendix D, "Getting Around a Template," to move the cell pointer to the cell listed in the instructions following the Goto label. With the cell pointer in the desired cell, type labels or numbers just as you would type them on a typewriter, and then press **Enter**. If you make any typos or other mistakes, see the box "Tip on Correcting Mistakes," on page 229.

- You can't move the cell pointer around the worksheet when a template menu is displayed (when the indicator reads *CMD*). To move the cell pointer, you first have to press **Q** (for *Quit-menu*) to remove the *CMD* indicator.
- Enter numbers using the number keys on the top row of the keyboard. Unless you are an experienced user, do not use the numeric keypad at the right end of the keyboard.
- Many of the cells in the template are protected. If you try to enter data in one of these cells, the computer beeps. If you are using the Lotus 1-2-3 program, this happens when you press **Enter** and you see the message *protected cell* in the lower-left corner of the screen. Press **Esc** to clear the message and continue.
- Many of the template's cells contain "formats" that affect the way numbers you enter are displayed. For example, you enter dollar amounts as .5 ($.50) or 200 ($200). The template automatically adds the dollar signs. You enter percentages as decimals, for example, .1 (10%) or .25 (25%). The program then displays them as whole numbers followed by a percent sign.
- When displayed, the *CALC* indicator on the screen indicates that recalculation is set to manual. Some templates have been set to manual recalculation because they are so large that recalculation takes a long time. If recalculation is set to automatic, the worksheet recalculates each time you enter data. When set to manual, it recalculates only when you press **Calc** (**F9** on the IBM PC).
- To recalculate the worksheet after you enter any numeric data, press **Calc** (**F9** on the IBM PC).
- To change recalculation to automatic, press **/WGRA**.

TIP ON CORRECTING MISTAKES

- To edit an entry as you make it (and before you press **Enter** to complete it), press **Backspace** to move the edit cursor to delete typos, and then resume typing. Press **Enter** to complete the entry.
- To edit an entry after you have pressed **Enter**, move the cell pointer to the cell, and then press **Edit** (**F2** on the IBM PC). The mode indicator reads *Edit*, and the contents of the cell appear on the second line of the control panel.
  - To insert characters, use the arrow keys to position the edit cursor, and then type in new characters.
  - To delete characters above the edit cursor, press **Del**; to delete characters to the cursor's left, press **Backspace**.
  - When you have finished, press **Enter** to complete the entry.

F. Using the Template Menus

The templates on the *Cost Accounting Resource Disk* have been partially automated for you. When you hold down **Alt** and press the letter **M**, a menu appears at the top of the 1-2-3 screen. This menu is similar to your program's command menu but has been developed specifically for the cost accounting templates on the *Cost Accounting Resource Disk*. This menu is called the template menu. When a template menu is displayed, an indicator on the screen appears to remind you. The indicator *CMD* appears on the top line of the Lotus 1-2-3 Release 1A screen and on the bottom row of Release 2 versions.

Template menus are used for entering and clearing data, printing templates, getting help, displaying graphs, or retrieving templates. A complete list of template menu choices is given in Table 16 along with a brief description of each. Although the choices vary from template to template, all template menus operate the same way.

- To display a template menu, hold down **Alt** and then press the letter **M** (for *Menu*). Throughout this book, this two-key command is abbreviated as **Alt-M**.
- To select a choice listed on a template menu, press the first letter or number in the choice's name or use → and ← to move the menu pointer over the desired choice, and then press **Enter**.
- To remove the template menu, use → and ← to move the menu pointer over the menu choice *Quit-menu,* and then press **Enter**. You can also leave the menu by pressing **Q** or by holding down **Ctrl** while you press **Break (Ctrl-Break)**.

Table 16
TEMPLATE MENU CHOICES (in alphabetical order)

Menu Choice	Description
AUTO123	Returns you to AUTO123 so that you can retrieve another template or quit 1-2-3. When you select this choice, a warning appears. • To continue (AUTO123 is retrieved and the current template is lost), press **Enter**. • To cancel the command and return to what you were doing, press **Ctrl-Break**. Then, press **Home** to return to the template and **Alt-M** to display the menu again.
Backup	Returns you to the previous menu if you decide not to make a choice from a submenu.
Clear	Removes the data you have entered and restores the template to its original state — just like it was when you first retrieved it. When you select this choice, a warning screen appears. • To continue (your data is erased), press **Enter**. • To cancel the command and return to what you were doing, press **Ctrl-Break**. Press **Home** to return to the template and **Alt-M** to display the menu again.
Compressed	Prints the templates in Exercises 5, 9, and 10 in compressed type on 8½-x-11-inch paper. Be sure the printer is on and has paper in it before selecting this choice.
Data/Cases	Displays a submenu of choices that automatically enter data into the template.
Graph(s)	Displays a graph if your computer is equipped to display graphics. If you select this choice, and your computer is not equipped to display graphs, the screen goes blank. Press **Esc** to return to the template. If there is more than one graph, this choice displays a submenu listing all graphs. When finished viewing a graph, press **Spacebar** to remove it and return to the Graph menu.
Help	Displays a list of cross references to the supplement *Applications in Cost Accounting with Lotus 1-2-3* so that you can quickly find topics that interest you. When finished, press **Enter** to return to the template.
Normal	Prints the templates in Exercises 5, 9, and 10 in normal type on 8½-x-11-inch paper. Be sure the printer is on and has paper in it before selecting this choice.

Table 16 (Cont.)

Print	Prints the template. Be sure the printer is on and has paper in it before selecting this choice. When printing the large templates in Exercises 5 (CASHFLOW), 9 (PRODCOST), and 10 (CAPBUDGT), selecting Print displays a submenu of print choices. You cannot use these choices unless you enter setup codes. This was done in Exercise 1, but if you did not do it there, you can press **S** (for *Setup*) from the Print submenu and enter them directly into each template. (You have to do this only once: they are saved in a file and used by other templates when required.)
Save	Saves the template, and any data you have entered, onto a disk using the original filename. You normally don't have to save the templates since they are already on the *Cost Accounting Resource Disk.* You save only when you have entered data you want to return to later.
Setup	When you select this choice, a screen appears, and you are asked to enter your setup codes for normal and compressed type. Type in each, and then press **Enter**. A discussion of these codes can be found in Step 5 of Exercise 1.
Quit-Menu	Removes the template menu from the screen and returns you to the worksheet so that you can move the cell pointer to enter labels and numbers.

G. The Function Keys

1-2-3 uses function keys for many important commands. Table 17 lists these keys and briefly explains the function of each.

Table 17
FUNCTION KEYS

Key	Function
Help [F1]*	Displays help screens.
Edit [F2]	Puts program into edit mode.
Name [F3]	In point mode, displays a list of range names.
ABS [F4]	In point mode, used to make cell references relative or absolute.
Goto [F5]	Used to move cell pointer to specific cell.
Window [F6]	When two windows are on the screen, moves cell pointer between them.
Query [F7]	Repeats the most recent data query commands.
Table [F8]	Repeats the most recent data table commands.
Calc [F9]	Effect depends on mode: • In *READY* mode, it recalculates all formulas in the worksheet. • In *VALUE* and *EDIT* modes, it can be used to convert formulas to their currently calculated values.
Graph [F10]	Draws most recently created graph on screen.

* The keys in square brackets following function key names are those used to execute the commands on IBM PC and compatibles. The keys assigned to perform these functions on your computer may be different.

H. Special Keys

1-2-3 assigns specific functions to many of the keys on the keyboard. Table 18 lists these keys and briefly describes the function of each.

Table 18
SPECIAL KEYS

Key	Effect When Pressed
Enter	Also called **Return**, completes entries, for instance, when entering data, answering prompts,or selecting menu items.
Shift	Enters uppercase letters and the upper symbols on other keys.
Caps-Lock	Toggles between uppercase and lowercase letters. It affects letter keys only. To enter upper symbols on keys, you hold down **Shift**. When **Caps-Lock** is engaged, an indicator appears on the screen.
Alt	When you have created macros, holding this key down while pressing the letter key to which the macro was assigned executes the macro.
Num-Lock	Toggles the number keypad on the IBM PC between being a cell pointer movement keypad and a number keypad. (You can also use **Shift** and the number keypad to enter numbers.) When **Num-Lock** is engaged, an indicator appears on the screen.
PrtSc	Sends whatever is displayed on the screen to the printer. (On nonenhanced keyboards, you have to hold down **Shift** while pressing **PrtSc.**)
Esc	When editing, it deletes everything from the edit line. When using menus, it can be used to back out by the same path you entered the menus.
Backspace	When typing, pressing this key erases the character to the left of the edit cursor. When pointing to ranges, pressing it returns the cell pointer to where pointing was started from and unhighlights the range.
Del	Erases a character over the edit cursor when an entry is on the edit line.

I. Using Fixed Titles

To look at templates that are wider than the screen, you press **Tab** and **Shift-Tab** to scroll the screen right and left. Unless column A is on the screen, however, the row labels scroll out of view. To keep row and column labels on the screen, use the /Worksheet Titles command.

This command cannot be executed when the template menu is displayed on the screen. (The indicator reads *CMD.*) If it is displayed, press **Q** (for *Quit-menu*) or **Ctrl-Break** until the mode indicator reads *READY.*

To fix titles, move the cell pointer to the row below, or the column to the right of, the rows and columns you want to fix. Press **/WT** (for *Worksheet*, *Titles*). Choices on the menu are described in Table 19.

Table 19
TITLES MENU CHOICES

Choice	Description
Both	Fixes the row(s) above and column(s) to the left of the cell pointer.
Horizontal	Fixes the row(s) above the cell pointer.
Vertical	Fixes the column(s) to the left of the cell pointer.
Clear	Removes fixed titles.

J. Using Windows

Windows, like fixed titles, are used to display distant portions of the template on the screen at the same time. This command divides the screen into two vertical or horizontal windows.

This command cannot be executed when a template menu is displayed on the screen (when the indicator reads *CMD*). If it is displayed, press **Q** (for *Quit-menu*) or **Ctrl-Break** until the mode indicator reads *READY*.

To create a window, move the cell pointer to the row you want to be the top row in the lower window, or the column you want to be the leftmost column in the left window, and then press **/WW** (for *Worksheet*, *Window*). Choices on the menu are described in Table 20.

Table 20
WINDOWS MENU CHOICES

Choice	Description
Horizontal	Row with cell pointer becomes first row in lower window.
Vertical	Column with cell pointer becomes first column in right window.
Sync	Windows scroll together.
Unsync	Windows scroll independently.
Clear	Returns display to a single window.

K. Quitting and Clearing the Screen

When you have finished with a template, you have two choices:

- You can use the template menu to return to AUTO123 to retrieve another template or quit the program.
- You can erase the template from the screen to create your own template.

To return to AUTO123

Press	**Alt-M** (if necessary) to display the template menu
Press	**A** (for *AUTO123*), and then press **Enter** (if you press **Ctrl-Break**) to cancel the command
press	**Home** to return to the template

To return to the operating system from AUTO123

Press **Q** (for *Quit*), and the prompt reads *No Yes*
Press **Y** (for Yes) to return to the operating system (or the Lotus 1-2-3 Access System menu)

To erase the worksheet

Press **Q** (for *Quit*), if necessary, to quit the template menu
Press **/WE** (for *Worksheet*, *Erase*), and the prompt reads *No Yes*
Press **Y** (for *Yes*) to clear the screen or **N** (for *No*) to cancel the command

L. Troubleshooting Problems

- If nothing else works, press **Esc** until the mode indicator on the status line reads *READY*. This will extract you from almost any problem and return you to a neutral state so that you can start over.
- If you make a mistake while pointing to create a formula, the mode indicator changes from *VALUE* to ***Edit***. When this happens, you will no longer be able to use cursor pointing. You have two choices at this point: Edit the formula to make it read correctly, or press the **Esc** key until the mode indicator reads *READY* to cancel the formula and then reenter it. (You may have to press it a number of times.)
- In many cases, selecting a menu choice calls up another menu. If you change your mind or can't find the command you want, there are two ways to get out of the menu:

  1. Press **Esc** to back yourself out of the menu one step at a time.
  2. Press **Ctrl-Break** to return to *READY* mode.

- Asterisks in cells (******) indicate that the number entered, or calculated by a formula in the cell, is too large to be displayed correctly. The number can be displayed by widening the column or changing the format.
- Printouts in compressed type when it wasn't requested happens when the previous printout was printed using compressed type. The printer remembers the setup code. To reset the printer to normal type, turn it off and then back on.
- If your screen goes blank when you select Graph from the template menu, your computer is not equipped to display graphics. Press **Esc** to return to the template.
- If the graph X-axis is not formatted properly, remember that required formats will change depending on the size of the numbers being plotted.